JAPAN
Akita
Aomori
HONSHU
Niigata
Sendai
Nagano
Tokyo
Yokohama
Mt. Fuji
Nagoya

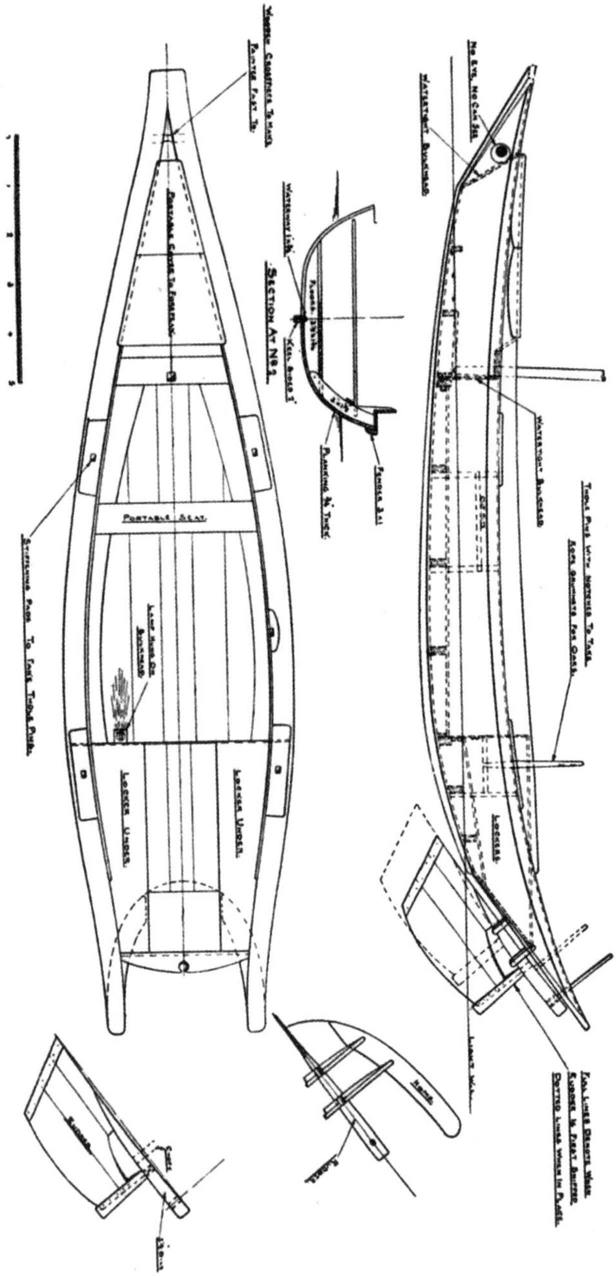

Image: Division of Work & Industry, National Museum of American History, Smithsonian Institution

Sampan Blockade

A SAILOR'S STORY OF SURVIVAL

by
WILLIAM C. WARWICK, Jr.

SlapDash Publishing, LLC
Carolina Beach, North Carolina

Sampan Blockade

A SAILOR'S STORY OF SURVIVAL

by

WILLIAM C. WARWICK, Jr.

Editors

Daniel Ray Norris, Amy Maria Norris

David Covington, Skipper Funderburg

slapdash • publishing LLC

311 Florida Avenue, Carolina Beach, NC 28428

910.232.0604 • info@carolinabeach.net

www.carolinabeach.net

LIBRARY OF CONGRESS CONTROL NUMBER: **2010924551**

William C. Warwick, Jr.

Sampan Blockade - A Sailor's Story of Survival

Carolina Beach, N.C., SlapDash Publishing, LLC.

220 pp.

International Standard Book Number: **978-0-9792431-7-2**

First Printing: **September 2010**

Additional image credits: iStockphoto.com - maps on p. i, iv-v and misc. cover art.

A True Life Story of Unrecorded History

Dedicated to my loving wife Elizabeth English Warwick for fifty-seven years of blissful marriage.

OTHER BOOKS BY WILLIAM C. WARWICK, JR.

A Sailor Comes Home

Changchun
Fushun
NORTH
SEA
Pyongyang
KOREA
Lüda
Kaesong
Inchon
Seoul
SOUTH KOREA
Taegu
Pusan
Mokpo
Strait
Hiroshima
Kobe
Korea
Fukuoka
Kita
Kyushu
Kure
Nagasaki
KYUSHU
SHIKOKU

ASIAN SAMPAN SKIFFS are small, flat bottom wooden boats of shallow draft. Their hull designs are about 15 feet or longer and are operated for transportation, fishing and shelter. They are stable work platforms used in coastal areas, bays, inland waters and rivers. They are typically open decked with a small cabin, but usually do not venture into rough water. Sampans are modified to suit needed conditions, the specific work for which they were built and are propelled by oars, sails or engines.

Common Sampan

Image: National Oceanic and Atmospheric Administration/Department of Commerce

Prologue

After many months of bitter fighting all over the Korean peninsula, United Nations (UN) forces had battled the communist invaders back to where they had started – north of the Yalu River which borders North Korea, Russia and northeastern China. UN field headquarters, assuming that total victory was just around the corner, released a field publication in which was written a promise to their freezing troops that we will all be home for Christmas. They wrongly concluded that the only thing left was a formal surrender. How wrong they were, because the worst was yet to come.

Back in China, their safe haven, the communist army, while waiting for the Yalu to freeze, was being replenished by Russia. Some argued that the frozen river could, after freezing solid, be used as if it were a highway. MacArthur did not believe that to be true, but time would prove him wrong.

During our enemy's rest and recuperation time in China, their strategy was to keep UN forces in a rigid defense posture so they had no time to learn that thousands of Chinese troops were being prepared to flank their military forces. But it is easy to understand why the UN generals didn't ever want the facts of that event to be known – it would likely expose a very serious intelligence blunder that had almost cost them the war. I learned about the Sampan flanking movement only after I had become involved with spying on that activity.

But by that time, untold numbers of enemy troops had already come ashore, flanking many of our troops. This flanking scheme by the communist army was so out of character for a primitive, peasant army that no one expected it could happen. UN war planners apparently wasted no time on such an impossible idea, even after being told by friendly North Korean fishermen that their Sampans were being confiscated in large numbers.

The enemy's plan was to remain visible and threatening while in China; therefore, with ordinary field binoculars they were in clear view while marshaling huge inventories of military hardware and troops. MacArthur became quite concerned and requested permission from U.S. President Truman to conduct massive bombing in China, which he believed would stop a second invasion by communist hordes. The UN and Truman denied his request because they did not believe the enemy would cross over the Yalu again, even after it had frozen solid.

The most dangerous thing about the communist flanking plan was that nobody believed sending small numbers of troops south in Sampans could possibly, with any measure of success, flank the entire UN command. Many people questioned being concerned about small vessels that could transport only three to five soldiers each. Anyway, they obviously **did** believe that a few little open-hull boats were of little threat to the huge UN military force, especially after considering the below-zero temperatures.

But they **were** a threat because **thousands** of Sampans would be used. Not withstanding the fact that Siberian Chinese and their Russian advisors are known for being the best cold-weather fighters in the world. Napoleon and Hitler learned that fact the hard way. Nevertheless, the flanking maneuver had already begun, but Truman and the UN still refused to allow any bombing in China, and that final decision became a very costly military mistake for the UN.

The communist army was about to score a perfect flanking of their opposition, one that offered no resistance and plenty of time, because no one on our side believed it could be done. The enemy wisely kept the UN forces on constant alert in order to pursue their real intention, which was a brilliant flanking deployment down the east coast of North Korea.

Historically, attempting to do the impossible has won countless military battles, and subsequently many wars. However, in the Korean War, every soldier, on both sides, spent most of his time trying to keep from freezing to death. But the communist fighters' big advantage was that they knew how to survive and fight in their frozen homeland.

A week or so before operation "Sampan Blockade" was authorized, our Ranger special operations team leader (Sergeant) received orders to evaluate all available intelligence and proceed to locate the coastal staging area for the claimed communist flanking maneuver. Before we left our secret operating hide-out in North Korea, code-named Nando Island, we hurriedly loaded our twenty-eight foot motor whale boat (MWB) with everything we thought would be needed.

Chapter 1

After two nights of motoring we had traveled north about two hundred miles, following the east coast of North Korea. Our destination was the delta of the Yalu River, which is approximately two hundred and fifty miles north of our Nando Island. With another night of motoring we should reach our goal, if we had not frozen to death before. As we traveled north it seemed to get colder by the minute.

On day two of hiding, I truly believed I was going to die unless I could find a way to get warm. Thank God I had not taken a sleeping pill because if I had, I surely would have frozen to death in my sleep. I thought moving about might help. But my body was so cold I had considerable difficulty getting up and out of my frozen hiding place.

Not yet completely standing, I smelled smoke from nearby burning wood and with joyful expectation began a very feeble walk toward the life-saving smell. Survival became first, above all other things, in my low-functioning brain, which was so near gone that I became very confused. Because of weakness, I crawled the last few yards to a small building which was the source of the smoke. Somehow I managed to strike the door. When it opened I was barely conscious and fell face down onto the icy snow.

Immediately two people dragged me into the heated little dwelling and quickly began removing my frozen clothing. I could hear but could not understand their Korean chatter while they were doing a number of things to warm my body and save my life. Very slowly my mind began to clear, and I took careful notice of the windowless little building, dimly lit by fire. It was all stone construction except for the small entrance door and low ceiling. I guessed it was no larger than fifteen feet by fifteen feet. There were several stacks of firewood but no furniture of any kind. The stone fire enclosure was located in the center of the structure's dirt floor, and the smoke was exhausted through a stone chimney straight up through the ceiling

The two people who had saved my life continued to observe me and rub my skin until they were satisfied that I should put on my warmed clothing. After I had done that, both of them stoked the fire, spoke some words I did not understand while tucking my body close to the heated stones, and then retired to their own warm spot on the opposite side of the stone firebox.

Feeling alone with my thoughts about my dreadful circumstances, the first thing that struck me was the awful stench. I would have run out of that filth the first minute if I could have done so. That example and many more are the basis of my opinion that North Koreans practice very poor hygiene. But I was alive and warm and for the moment that was a powerful reminder of the debt I owed to those two unwashed North Koreans who had without a doubt saved my life. My mind must be clear before I do anything, I thought. First, I must determine whether or not it's still daylight. If it is, I'll wait until dark. Further rest and warming will serve me well anyway. Since it's probably snowing outside and people are living in this little refuge, military boot tracks are of no concern.

Subsequently, one of my rescuers got up from his/her bed, walked to the door, removed the small piece of wood that kept it closed, opened it, went outside and quickly closed it. In less than two minutes he/she was back inside and returned to their warm spot. A normal biological activity had become my ticket to freedom – that is, if they accepted my need as well. Yes, that's it. Someone went outside to urinate, and I had learned how to manage the door by watching the previous user. If only I can quietly do the same thing, I thought. But I had also learned through the open door that it was dark, and that really up-set me.

All I could think about was what my failure to return had done to the team. After quietly standing up to evaluate my strength and snug-up my clothing, I decided to leave immediately. Then I discovered that my sidearm was missing, which was a huge problem. If the enemy got their hands on that US-stamped pistol, they would be alerted that American spies were in their country and that would start a massive search. I must retrieve that weapon because the Sergeant will kill me even if the North Koreans don't.

I took a few quiet steps toward the opposite side and dimly spied my holstered .45 caliber handgun and belt on the dirt floor against the stone wall, uncomfortably close to the snoring sleepers. God you have brought me this far please don't fail me now, I prayed, while reaching for my weapon. Got it! Thank you God. Now please keep them in dreamland, until I'm gone, so I won't be forced by survival circumstances to kill two kind people who have just saved my life. While putting on my gun belt I began carefully moving toward the door, taking notice that the sleepers had not been disturbed. There was no problem exiting the warm little refuge.

I gently closed the crawl-through door, placed a stone against it to keep out the arctic air, and whispered, "Goodbye kind people, thank you for saving my life. May God protect you always as He has this night." Then I hurried away from the small, primitive, stone building.

Chapter 2

While carefully walking fast through the darkness toward our rendezvous area, I felt very thankful that it had not been necessary for me to kill those two harmless peasants. I am not a true warrior – I thought, therefore killing people, or anything, without good cause is not my cup of tea.

Suddenly I heard "Nando" and quickly responded in like kind. Again and again I heard our password and again and again I responded. My anxiously waiting teammates had formed a large half-circle from the boat to about one mile inland, to ensure a better chance for hearing my password and for me to be better heard by them.

They had good reason to suspect that I had been either killed or captured. I had been captured, but in a very unusual way.

Everyone, including our leader, was obviously overjoyed at my reappearance. An expected reprimand from the Sergeant didn't happen. My delay had given all of them quite a scare. It had been more than an hour past the time we should have left.

The Sergeant, being the cautious man that he was, began rapid-fire questions. He was satisfied with all of my answers except why I had not killed the two North Korean peasants. Again I answered that killing them with my sidearm would have been too noisy and that would have surely set off barking dogs. Dead bodies, when found, would also cause big trouble for us, so I left them sleeping. In my weakened condition the two of them might have been too much for me to physically subdue. So using any known physical attempt to do them in was ruled out.

Yet another consideration was that their multi-layered clothing made it impossible for me to know if the two who had saved my life were young or old, male or female, weak or strong. That left only my sidearm and again, that choice would have more than likely been fatal for all of us. Anyway, they had already demonstrated no interest in me beyond saving my life and believe me, they did that with skills learned from living in a cold, primitive environment. Finally, I fully believed I had done what I had been trained to do ... **think**.

At the end of the interrogation the Sergeant bluntly said, "Show me your cyanide pill." I reached into my clothing where I had previously concealed it and thank God it was still there.

He smiled and said, "I'm very proud of you, Ranger. Everyone, including me, is very relieved that we didn't lose our boat captain ...." (pause) "Take us to Vladivostok, Captain!"

"That I will, Sergeant!"

"Sam, is our boat topped off with fuel and ready for sea?"

Sam answered from his seat adjacent to the engine, "Yes sir, Captain!"

"Start her engine," I ordered.

"Aye, aye Captain," he answered.

After getting underway, I recommended to our leader a course that would shorten the distance to Russia by thirty to forty miles and thus save time and fuel. "You have my permission to do that Captain," the Sergeant replied.

My navigation strategy was to avoid looping a very large cove by steering northeast (away from the coast) for about twenty miles, and then steer northwest to a position which was only a few miles south of the Yalu delta.

There are two range light towers in a huge bay, that have twenty-five miles visibility on a clear night. They are used to guide surface vessels into Vladivostok's busy port facilities. I had read about them earlier while studying my charts and navigation data. From a landing in northeastern Korea south of those lights, we should be able to see Vladivostok's night lights and then make further landing decisions.

I reminded Sam that we must reach the coast before daylight or the Russians would be all over us.

He asked, "How much time do we have?"

"Seven hours max," I answered.

"What's the distance?"

"Sixty to seventy nautical miles."

"How much help can I expect from the sea?"

"None until I turn northwest in an hour or so, and then maybe one to two knots [2 to 3 MPH]."

We had done everything possible to protect our boat from radar but an overworked engine's failure would be the ultimate disaster, and that concerned me deeply. It seemed to be a good time to recommend to our leader that the boat's large white cover be used as a wind break. "Where is it located?" he asked.

"In the storage area beneath where I'm sitting. The Navy crew must remain on duty so we can't help, but use the cover as you wish."

Surprisingly, the Rangers made in a rather short time a very efficient shelter which shielded us from the freezing wind and the cold sea spray. "What a blessing that is!" I thought.

When they had completed the much-appreciated and useful wind screen, the Sergeant approached me and said, "Up north in an all snow-white terrain we can make good use of the white cover to camouflage our boat." I agreed with his creative thinking. The finished shelter was blocking my forward view but no matter, the sky was my compass. I will just lay back and steer by one star among countless others. Anyway my indispensable bowman was at his duty station looking and listening for anything unusual and that was very comforting for me to know.

Sam, our mechanical genius, was busy keeping watch over our 120 HP Gray Marine Diesel engine which was, because of the long haul and overload, beginning to vibrate and make other sounds which more than a little concerned our engineer.

We had little time, before leaving Nando Island, to think about what our needs might be to keep us warm for a week in the frozen north near Vladivostok. Consequently, a hurry-up mindset developed while loading the boat, which was done without any consideration of balance. Actually we had loaded far too much cargo, because we had loaded things we would not need, such as too much diesel fuel, which later would prove to be a blessing, and did not load enough things that we would need, such as blankets and enough clothing to keep us from freezing to death.

During the first mile of our north-bound mission, I knew from unusual engine noises and slow throttle responses that we had clearly overloaded our boat. But for such a long trip I wrongly guessed that we would eventually need everything on board. In the beginning, however, I was only a little concerned and did not think I should bother our team leader because he was obviously excited to have been chosen for his new assignment and was therefore anxious to move out as soon as darkness permitted.

My regrets came much later when trouble caused me to pay the consequences of careless preparation. All I could do to correct or even improve the problems was to move the boat's center of gravity aft (toward the steering cockpit) and trust that would raise the bow a few inches and reduce resistance pressure on the dangerously low, plowing bow.

As we were, all it would take to make us frozen history would be for just one four-foot wave to come over our bow and flood the pumps and the boat. By raising the bow perhaps four to six inches it might reduce resistance enough for her to knife through the water instead of plow.

We could not throw anything overboard to lighten the load because if any piece of the castoff was found by our enemy, it would telegraph our presence. The Sergeant would certainly shoot anyone who made that mistake.

We had with little warning begun to experience strong headwinds. Sam and I decided it would be wise to execute my intended northwest turn right away and hope for safer following seas. We agreed that it would be suicide to continue steering our overloaded boat into increasing headwinds that were parenting three- to four-foot seas.

Our plan was to slowly bring her bow to a northwest heading without changing the engine's RPMs. We decided to let our boat find her own way to optimum coursing so that we would know from her adjustment to sea conditions whether reduced power was the thing to do, or if a manageable downwind course could be found that would lesson our danger. We very slowly and cautiously began turning her northwest while praying a rogue wave would not come over her bow and sink us.

It seemed an eternity, but finally she settled down to a much quieter running engine with no vibrations, and a bow with greatly reduced plowing. She had found her way and was coursing as well as could be expected before a two-knot (estimated) following sea. After things were right for her, the engine began purring like a kitten.

Chapter 3

Only Sam and I knew how close we had come to losing our boat. The early turn was without a doubt a must. After a lengthy conversation we (Sergeant and I) decided to put our trust in the two range tower lights which should, if I was right, begin to peep over the western horizon in one to two hours.

Considering everything, twenty-five miles was a safe distance before attempting a night landing in what seemed to be the beginning of a blizzard. At least from the range tower lights we would know exactly where we are, because they are stationary 180 degree lights, ranging true south when aligned, and that is as good as it gets for any navigator to get a fix and know where he is. To verify a position is truly like finding heaven's gate. Those two Russian range tower lights will be, when we see them, like the gift of life.

There are seven men depending upon my seamanship, and that concerned me very much. But for the time being, since our boat was surfing fast and safely, I resolved to continue steering her downwind, which is an easy course – because if the man at the tiller gets inattentive and wanders off course, even a little bit, from where the vessel wants to be, she will react violently to remind him to get back on her optimum course.

When I see those range tower lights, only then will I consider a course change. But while waiting, minutes seemed to have been hours. If I'm wrong about those lights, we are all dead men. But I'm not wrong because I memorized all of my charts of this whole area.

Maybe I should think of something pleasant to help the time go by. Something like the shelter. Yes, the shelter, that's a good thing to think about because it has sure served us well. The two previous nights, all of us nearly froze to death, but not tonight. Just being protected from the wind seems to have cut the cold in half.

I played several other mental games and estimated passed time, over and over. God, seven people have put their lives in my care. Please help me get them ashore safely, I prayed. Believe it or not, He spoke very softly to me and said, "My son your prayer will be as you have asked. Look now at the western horizon."

At that exact moment I heard my lookout practically screaming, "Captain! I see lights, low on the horizon two points off the starboard bow. Lights, lights, I see lights! It's real, look two points off the starboard bow (west)."

While the lookout was still shouting his sighting I stood up in the cockpit so I could see over our shelter, and there they were.

"Yes, I see them!" I shouted back to the lookout. God had answered my prayer.

"On deck, Sergeant! I have something to show you!"

He was out of the shelter in a flash repeating, "What is it Captain?"

"Look there," I encouraged while pointing and added, "You are looking at the range tower lights from Vladivostok, Russia." All of us were overflowing with joy.

The back (tallest) green range lights had just begun to appear. Every one cheered the bowman for his superb vigilance. He was even less talkative than Sam but when he did talk it was always something you needed to hear. One of the Rangers spoke up and said, "I thought I'd never see lights again."

"Do you know where we are, Captain?" the Sergeant asked.

I welcomed that question and proudly answered, "That I do sir, that I do. We are close to thirty miles south-southeast of Vladivostok, Russia, and roughly the same distance nouth-northwest of the Yalu River delta which is the border separating North Korea from Russia. The tallest range tower lights are barely visible above the horizon showing a true bearing of south-south-east. Not yet being able to see both range tower lights tells me that it is close to thirty miles distant from us. That is how I know, within five to ten miles, where we are."

"Coxswain, how in the world am I supposed to give you a celestial star to follow that will top that?" All of us got a real big laugh.

"Sergeant, just give me a course south, I'm on a roll and I can find the Yalu's delta after we've motored south an hour and then turn west. There's one thing more you should know, darkness is running out. My estimate is that we have little more than two hours of darkness left."

"How do you know that?" our leader asked.

"Well, what we are seeing now are the highest tower lights, the lower tower lights will not appear until we are about ten miles closer. However, the lights on both towers will be turned on from dusk till dawn only, which is good for us. Because the dusk-till-dawn rule requires red to be turned on two hours before dawn. When red replaces green that means there are two hours remaining for ships already in the twenty-five mile channel to reach Vladivostok.

"Last but not least, we do not know how long it will be before the green lights we are now looking at will change to red. Perhaps more than a few minutes can be gained from that event. So at worst we have two hours of darkness left. At best three hours. But there is one thing we can count on, Vladivostok is a top secret Russian navy base and intruders like us will be shot on sight. Nothing but Russian ships and submarines are permitted to enter this harbor. And you can bet they have very sophisticated detection devices all over the place. That's why things here are different from any other international harbor entrance. Sergeant, I recommend that we head south immediately."

"See that little star approximately 45 degrees above the southern horizon?"

"Yes I do, but I'll point to it with the bow so that I can be sure. Okay, I'm on it Sergeant, check it out."

"That's it; your heading now is south!"

"Good, I need to stay on this course for one hour and then make a 90 degree starboard turn."

"Call me if necessary."

"I will, thanks, Sergeant."

My mind drifted to the Sergeant's inflexible will to succeed, which included, "Do not lose this boat because of being visible on the water during daylight hours!" He really would kill anyone who would carelessly put our team in jeopardy, of that I was sure. I remember his often saying, "I expect the best from each of you all the time and each of you can rightfully expect the same from me."

While thinking of my responsibility to my teammates, I realized that I must have the boat ashore and camouflaged before daylight. And believe me, I prayed hard for God's help, because this time there would be no huge range tower lights to help me. The best I could hope for would be a small lighted buoy at the mouth of the Yalu with no more than a three mile visibility. What color light is anybody's guess.

Chapter 4

I leaned forward and tapped Sam's shoulder, and he turned and leaned in my direction so we could talk. "Sam, as you probably know I am very concerned about reaching land before daylight, therefore I need every turn of the propeller you can safely give me. Please remember what I have said is a request not an order."

"Bill, I have been pushing her hard since we started south, so far everything sounds okay. I'm sure you know that when we turn to starboard we will be running before a pretty strong gale, and with that help I will get about twelve knots."

"I hope so. But this storm has been increasing in intensity for several hours and I'm beginning to worry about that. We will consider the weather again before I make the turn. But in the meantime tell the Sergeant I need a conference with him."

Almost in the blink of an eye he was in the cockpit.

"Sergeant, I'm sure that you have observed that the weather is crashing and I've decided to order 'condition red' which is a standard U.S. Navy survival-at-sea drill. Trust me, your Navy boat crew is very experienced and highly trained in this sort of thing. However, we will need all hands to help because we are overloaded and this storm is developing fast into a blizzard. What I want you to do is report to me exactly what is happening in the boat at all times so that I can make tiller adjustments before it's too late. Keep a man between you and me for an instant communication link.

"When and after we turn, the big issue will be taking on water, which will change the boat's center of gravity in a heartbeat and pitchpole, which means to flip her upside down. One man should be assigned to the manual water pump at all times. That same person should also monitor the automatic pump in the bilge. The bilge must NOT flood and I'm asking you to make sure that does not happen. But you cannot do what I'm asking of you without light. Sam has two battery/DC battle lanterns stowed under his seat – get them and use them.

"Now, as you have so well taught me, NO mistakes if you want to live. The Engineer and I cannot leave our seats for one second because we must totally concentrate on our battle with the storm. What we are doing and what the others are doing must be coordinated. Absolutely everything must be securely battened (tied) down and that includes me if you want me to remain in this cockpit.

"Questions?"

"Not at this time Captain."

"Please send the lookout to the cockpit."

"Will do Captain."

When he arrived, even a dim look revealed what the freezing weather had done to his face. It was shocking but there was no time to deal with that. With his experience he could have advised me about the weather!

I told him to make his bow peephole watertight with the boat cover and then to do the same everywhere. "You know as well as I what will happen if we get too much water in the bilge. We will end up in the drink where our life expectancy will be three minutes. But hear this, I had rather have you and Sam with me during this extremity, than any two sailors in the whole wide world.

"Check the bilge pumps first, I hope they have not iced up. If they have, bailing is the only thing left. Use the Rangers' helmets if necessary. One last thing, I'm going to attempt a down-weather starboard turn as soon as the boat cover is as watertight as it can be, so let me know."

"I will Captain."

"Also I want you to deploy the sea anchor when I call for it."

"I'll get it ready for use now."

"Good man!" I replied.

For those readers who do not know anything about a sea anchor, maybe the following explanation will help. Its primary purpose is to control a boat's speed when running before a storm. It also prevents pitchpole, which means to dive into a trough and never come up. When deployed under water the sea anchor appears as a small parachute being dragged by the boat.

Preparations for the dangerous starboard turn were nearly completed. My bow man was ready and waiting for my order to deploy the sea anchor at just the right moment, without which we had no chance at all. And I was anxiously waiting to hear from the Sergeant about his men securing the cargo, checking the bilge (bottom of boat) for water and doing whatever else was necessary to keep water out of the boat during the turn.

Our lives were hanging by a thread and I would not turn until I had done everything possible to improve our chances of staying alive. I leaned forward and informed Sam, who was always ready, that when the right wave came along and the bow began to rise enough for my satisfaction, "I will shout 'STAND BY'. And if conditions are right at its crest I will shout 'EXECUTE!'"

He already knew that at that instant the sea anchor will be deployed and I will jam her rudder to full starboard for only seconds before pulling back to neutral hoping, to make a 90 degree turn on top of the wave instead of diving in front of it, which would submerge the heavy bow and would surely be a fatal mistake.

Simultaneously, Sam will begin his unending effort to adjust the boat's speed to equal that of each wave so that it will surf and not dive. Too slow is deadly because the stern-deck will flood with faster waves breaking over and into the cockpit and too fast is likewise deadly because the bow will, without a doubt, dive under a rogue wave and stay there.

I must continuously adjust the tiller to Sam's speed choices and steer a perfect down-weather course. What Sam and I were about to do can be done successfully if we remain perfectly coordinated, but if we do not, success will be very unlikely. Even though Sam and I had practiced the upcoming exercise several times before, nothing can ever be taken for granted in a storm at sea.

I know what we were about to do will be strange reading for some. However, it's a standard U.S. Navy survival-at-sea drill. To this day I am still a certified U.S. Navy lifeboat captain.

The reader might question my ability to, without any lighted instruments, steer my boat in such terrible circumstances. My large tiller was made of hard, solid brass and through it I could feel all kinds of vibrations from the boat and from the sea, good and not so good. For better understanding, a good example might be an automobile's reaction when it has been driven off smooth asphalt. Its bouncing and bumping is telling the driver that he/she is off course. Lights or not, a boat in trouble will tell the coxswain the same thing and he had best listen and know what to do.

Finally, the Sergeant approached me and said, "Captain, both pumps are working as they should, the bilge is dry and the boat cover is as watertight as we can make it. Is there anything further we can do?"

"YES! Hold on and pray because in about one minute we will know whether or not God has answered our prayers."

Two or three big waves lifted the boat but did not raise her bow enough and then as if sent from heaven, I felt the boat being raised bow up about 10 degrees, just as I had prayed for.

"STAND BY!" I shouted.

The rising wave felt better and better, I knew in an instant that God had provided the prayed-for wave.

"EXECUTE!" I screamed, and instantly jammed the rudder hard-over to starboard fully aware that I must pull it back to mid-ship in seconds or she would broach (roll over) and sink. Within those seconds I felt from the tiller reduced vibration, which made me know that we were on top of the large wave and that Sam had cut back engine power.

He knew by its laboring sound that the sea anchor had engaged and was protecting us from a pitchpole (end-over-end tumble). I knew all of those things, but because of the stormy darkness and no navigation lights, we couldn't see ten feet. I also knew God was with us but I couldn't see Him either.

I had put the tiller back to mid-ship and our boat was on a downwind course to who knows where. Our leader came from under the protective cover and excitedly said, "Less than a minute ago, I thought by now we would all be frozen to death. No other boat crew in the world could have done what my crew has done! And would you believe that the bilge is still dry?" That really pleased me because there was still a blizzard pounding away at us with snow, sleet, and freezing wind.

But without the waterproof boat cover (shelter) we would not have survived, because we did not need to get wet in below zero weather. The Rangers were still trying to capture the heat from the engine. Being inventive as a consequence of their training, they were working hard trying to stop the freezing weather from entering the shelter they had constructed from the boat cover. But since the wind was so terribly strong it was no easy task.

They finally realized that I was the only one who was not protected from the arctic wind, and immediately began a host of strategies that would hopefully keep me from freezing to death. They offered gloves, which were refused. "You could lose your hands if they freeze."

My response was, "But that would be far better than losing the boat and consequently all of us. Anyway, my hands have already frozen to the brass tiller, but I can still feel storm messages from our boat and then know what I should do."

I would not give up the tiller because I was the only one on the struggling vessel who had the experience to cope with such an angry sea. The Sergeant asked for my permission to wrap the exposed metal on the tiller, and also the top of my naked hands. And then he sternly reminded me that if I refused,

I would soon lose all feeling in both hands. Because of that grim reminder, I agreed but insisted that they must always leave my arms unrestricted so that I could make quick steering adjustments.

I could not recall how many times in the past that I had been totally amazed at the versatility of our Rangers. Where does our country get such men? I thought.

When they had finished fixing me up using battle lanterns for light in sub-freezing weather, with only their medical supplies and talent, one might have thought I had been to a plush hospital. They cut arm and leg holes in several body bags and used them to insulate my torso. The bags were zipped up to my eyes which I didn't need at that time anyway because it was dark and I was steering by what I felt from the tiller. But they made a significant improvement in my exposure to cold and suffering. God only knows what else they used to protect my extremities. In any case their initiative and their creativity had certainly saved my life.

By the way, I refused all medication because my sense of feeling was all that was keeping us alive. Almost freezing to death twice in one day – I wondered if that were some sort of record or had I without realizing it just carelessly underestimated winter at that latitude.

After doing all they could for me, my teammates sought refuge in the cabin. However, I asked the Sergeant to stay with me and hear my strategy talk.

"Sergeant, it's my opinion that this storm will probably last two or three days. If that is true, the worst is yet to come and I'm about to tell you why I believe that. This is absolutely the worst storm I have ever encountered in a U.S. Navy whale boat, which is the most seaworthy open-hull watercraft in the world. I do not know our current position or exactly how long it will be before we will be driven by this raging storm onto a rock-infested shore which will destroy our boat and us.

"But I do have a plan. The following is a must if we have any chance at all to survive this deadly situation. Here it is. I told you earlier that I had memorized the charts of the entire area surrounding the Yalu River delta. I clearly recall there are many small islands about five miles, more or less, offshore from where that river flows into the Sea of Japan. My memory also tells me those small and large rock islands extend north and south along the North Korean and Russian coasts for a considerable distance. If we are lucky, the wind and waves will carry us to one of those rocks, but it will probably be daylight before we reach the delta area. Hopefully, even in restricted visibility, we will be able to see one close enough and big enough for me to execute my plan, which is to steer to the lee (calm) side of one of those little islands and hold up until the storm passes. Although it may prove to be impossible, because maneuvering a small craft on this stormy sea is next to begging for a rollover. But this time we have no choice!

"Here's how I plan to attempt our last and only chance. Every man will be assigned a specific task, which will be practiced and timed to dovetail with all of the other team members, because it must be done super-quick with no mistakes. We'll get only one chance before we are either dashed ashore on the rocks, which will destroy our boat, or have a rollover in arctic seawater, and freeze to death.

"Sergeant, I want you to act as my communication link with everyone, but especially with our engineer. I repeat, we will get only one chance and there is no doubt that we are already within an hour of crashing ashore somewhere. When that happens we will have lost the only thing (our boat) that has been keeping us alive. Sam and I cannot take our attention away from the boat for one second, so finding an island will be up to the rest of the team. I repeat again, if we fail to find an island we are dead! This boat cannot survive coming about 180 degrees and then heading up into eight- to ten-foot seas for more than a few minutes. However, if we find a rock, Sam and I will desperately try to get us into its lee (no wind). To be perfectly honest with you, I believe that if we do not get into the lee of land somewhere within the next hour, we will be slammed ashore with no hope of living more than a few minutes. But we do have a chance, even though a slim one.

"At this time, I do not know in which direction I will turn, but if an island is sighted, even though dangerously close to shore, I must go for it and pray that I make it. The first thing in my plan that must be done after an island has been located, if that happens, is to release the sea anchor line at the exact instant I give the order to do it. Anything less will be fatal. Yes, I mean release the sea anchor line, because retrieving it will take too much time, which we do not have, and in a storm like this it is too risky to haul in anything that could get close enough to be caught up in the boat's propeller and break the shaft. Anyway, I can make something later to simulate drag if it becomes necessary.

"What I must do, if I'm to have any hope of success, with a reverse course (180 degree) maneuver in this blizzard, is to release the sea anchor and simultaneously begin the fast turn. If successful on top of a wave, she will slide into its trough with my holding the rudder at approximately 15 degrees off center. And then if we are still afloat, I will prepare to meet the next big wave in the same manner.

"The turn must begin no more than one second after the sea anchor line has been cast off, because without its drag on the stem to keep her from dancing, I would in the blink of an eye lose control and our boat will roll over. Sergeant, the bow man is trained to use a special kind of knot on the cleat that's holding the sea anchor to the boat, therefore he should be the one to release it. One man should stand-by the pumps at all times and the remaining three have perhaps the most important responsibility of all, they must keep water out of the boat's bilge during the very fast turn. So I recommend that you inform and prepare our team and get back to me quickly if you need further clarification or time.

"Sergeant, if we don't make it, I want you to know that it has been a rare privilege for me to have served under your command. It sure beats getting shot! Or does it?!"

We both had a big laugh before he spoke up and said, "Thank you Captain but as long as you are able to manage that tiller you are frozen to, I believe we will make it."

"One more thing, Sergeant. Tell the bow man I want him to stand his lookout watch here with me, now! Because his sighting skills are necessary for my decision to reverse course which I will do the instant the sea anchor line has been cast off."

"As the Navy says, aye, aye Captain!" A dark, gray dawn had begun to appear and the lookout reported that fact to me.

"Good!" I responded, "but find me a rock to hide behind, SOON!"

Chapter 5

Monster waves that were about to kill us were at that time foremost on my mind. Since I was upwind, useful surf noises from ten-foot seas pounding the coast couldn't be heard – therefore I could run aground without any warning. Deprived of any navigation aids, I was getting more and more concerned about rocks coming through the bilge planking at any moment.

During my uneasy thinking, the lookout shouted, "Captain, an island, I see an island!"

"Where away?" I loudly inquired.

"Off the port beam."

"Give me an approximate range."

That was a very difficult request but he had eyes like a hawk. "Fifty to sixty yards."

God help us, I must turn this boat NOW, I thought.

"Okay I can barely see it, stand-by to release the sea anchor. Sergeant, I have a target on our port beam and I'm going for it, because the waves are increasing in height which tells me that we are in dangerously shallow water and are about to run out of time. Tell our engineer to give me all the speed he's got when I begin my 180 degree port turn. I'm going to run as fast as possible to the lee of our target, if I can keep it in sight."

When a wave lift felt good, I shouted "Stand-by."

"Execute!" followed after I had coordinated my order to release the sea anchor with my feel that the boat was absolutely on top of a wave.

When our heavy boat started down the weather side of a huge wave I pulled her rudder to about 15 degrees off center to begin sideslipping, which saved our lives – of that I'm sure. However, anything was better than diving an overloaded boat into a 10-foot-deep trough using neutral rudder.

The bow didn't go under on the first try and after I realized that wonderful godsend, I vowed to use the same strategy with every wave. That decision served us quite well again and again as I ran full speed in search of the calm water on the lee of the ghostly island.

Sideslipping took more time to reach my goal but it finally happened with all of my teammates still safe. Finally, by the grace of God, the blizzard's powerful wind side slipped us abeam of our target. Sam carefully reduced her speed until the target was close off our starboard bow.

The rest was easy. Starboard the tiller and run like a bat out of hell for calm water behind the target, if I could find it. And believe me, I continuously prayed that I would find it. Running a boat full speed across water which has unknown hazards and, if the hull had been ruptured, would have quickly killed all of us in minutes. That was no easy decision to make. But it was a chance for life, and if God leads us to calm water behind a rock that I could barely see, we will live. If not we will die.

The tempest seemed to be winning, but there would be no turning back. I became more and more determined to have the last word about who was winning. Taking very heavy weather hits on the port side made steering a nightmare and it was only through God's divine blessing that I was able to continue. And for that I was very thankful because as a U.S. Navy boat captain, I always held a deep sense of responsibility for the safety of my passengers.

Suddenly, while pointing the bow up into the storm a bit to reduce the chance of being washed ashore, steering became a little easier. Is it possible, I wondered, that I'm moving into the lee of the target? But not being certain due to the very handicapping visibility, I continued the fight to keep our vessel afloat in a terrible storm, something that was becoming dangerously close to a rollover.

I began to lightly feel, through the tiller, that the blizzard was easing off – namely, it was not as ferociously trying to throw me out of the cockpit. But the reason that did not happen was that the Rangers (people who think of everything) had tied me down to stay in the cockpit, and I do mean to stay. To get careless now would be fatal for all of us, and that I would not do, I vowed. But the boat continuously became quieter and easier to manage. No more pounding and bouncing all over the place.

After waiting a few moments for the miracle to sink in, I shouted over and over, "We've done it!"

The sea had held us hostage for many hours, and now this wind-protected little refuge was as quiet as a mill pond. We had found what we so desperately fought and prayed for!

"Sam, you and the lookout will have trouble finding a proper place to anchor because of poor visibility, but I must leave the task of anchoring up to my Navy teammates because I do not believe I can stand up. Don't forget to use the lead-line often and keep our depth at 10 feet or more. Also let her swing on about 100 feet of anchor rode (2-inch rope) and keep an anchor watch at all time in case she drags or hits something submerged as she swings on the rode."

I know it is unbelievable but my pain had changed to a stupor in which I was barely conscious. I was only slightly aware of my teammates who were all over me doing this and that. One thing I do recall though was how systematic the Rangers were. They had already learned how to detach the tiller from the rudder post, knowing that would be immediately necessary so that I could be quickly moved to a warm place adjacent to the engine. I have no memory of being moved.

The Rangers were all superbly trained emergency medical people, always prepared! They had fixed a bed for me close by the overworked and hot diesel engine, where they took off my frozen clothing and began the slow process of warming my body. Yes, I had almost frozen to death again. This time the boat was anchored, so the wind was blowing bow to stern. The makeshift cabin that the resourceful Rangers had made could now be heated a little with our boat's engine.

I recall hearing Sam report to the Sergeant, "There is plenty of fuel, I'll idle the engine for as long as it takes."

I was given pain medication and know very little about freeing the tiller from my hands, but I was later told it took a long time. And guess what, I was also told the Rangers had used Sam's two battle lanterns for light while working on me. Anyway, who but a fool would be out in a blizzard looking for spies?

In the late afternoon I began coming around and asked for help to sit up. My hands were wrapped like boxing gloves. "Hey fellows, I've got a problem here!" Everyone had a big laugh but the Rangers had already arranged facilities for a man with no hands. I will leave the particulars of their arrangement to my reader's imagination. The Rangers were constantly hydrating me with warm water. And let's not forget K-rations which are only one step above starvation. However, Hershey's chocolate candy bars were plentiful and were a wonderful way to end a K-ration meal.

The worst of the blizzard seemed to have blown over so I asked for help getting to my cockpit. My crew helped the Rangers bundle me up warm and even though it was still snowing, seated me where I had so recently, with God as my co-coxswain, survived the most intense storm at sea of my career. "Thank you mates", I said. "Now I invite you to join me in a prayer thanking God for His gift of life."

After our prayer, I announced, "As of this moment I shall with God's help and our leader's permission continue to be your captain."

Everyone cheered loudly! The Sergeant called for silence and reminded us that we were spies and must not help our enemy locate us. I continued, "We are blessed with having the bravest, most talented group of men that have ever come together for any purpose. And it is my firm belief that we will successfully complete what we came north to do." More muffled cheers. "And to do that we have a leader who knows how to get it done." More cheers and another call to silence.

"While we are waiting for this blizzard to pass, I will recommend to our leader that we give this boat a thorough inspection and if any part of it has been compromised, I must know where and what. The first thing I intend to do is make a replacement for the sea anchor I sacrificed to the Sea of Japan."

"No you won't!" our leader cautioned. "Not until I check those hands. Put him back in his bed."

After returning to my warm bed by the engine, the Sergeant began with extreme caution removing my bandages. I tried to be brave, however I soon lost that pretense because it really hurt. His first expression was what every medical student must learn on his first day of training. "Hmm … improving but a long way to go. No signs of any infection yet. Now, I want you to clearly understand what I am about to say, this team needs your brain not your brawn. If you can accept that fact and protect your hands totally for a few days and if no infection develops, I will allow you to stay with the team. But if infection happens, I will have you evacuated as soon as possible. A last reminder – if you fail to follow my instructions, including medication, you will become infected."

"I promise to follow your instructions unless …."

He cut me off. "There will be no unless!"

"All right Sergeant, no 'unless.' "

Our team leader assured me that our boat would be thoroughly checked for unseen damage. My problem was that I just did not like being idle. Sam was close by and we chatted a little while he was servicing his engine.

"Any serious problems Sam?"

"None that I can't fix but it is going to take time, I'm charging the battle lanterns now because I'll need them when I shut down the engine for about three hours. I have all of the tools and parts I'll need."

The Sergeant was working on the entrance to the deck house when I got his attention and signaled for him to approach me.

"Sergeant we have a problem that will require shutting down the engine for approximately three hours. But we should conceal our boat before we do that because hanging on an anchor rode would leave us helpless and visible if it suddenly clears up."

"Great idea coxswain, just tell me what you want to do."

"First make sure the tiller security pin that holds it to the rudder post is properly engaged. Next, fix a place for me to stand in the cockpit, even if someone needs to hold me up. Sam, can we move the boat?"

"Yes, any time you are ready Captain."

"When I'm ready, move the boat forward as the rode is hauled in until she is directly over the anchor, and then go two bells (neutral) until the rode and the anchor are on deck."

"Aye, aye Captain!"

"Remember we must keep the propeller clear of the rode or we will tangle up and break the shaft."

Two Rangers were assigned to carry me to the cockpit, stand me up and stay with me. After they had done that, I told Sam that speed would be his choice. I selected the Sergeant to man the tiller. He was quite reluctant but I told him that I needed him and he agreed. Also I advised him to "Just listen to what I say after I speak the word "tiller."'

I put our lookout and two Rangers forward on the anchor rode and cautioned them not to leave any slack rode in the water.

"Tiller, neutral. It's your boat, Sam!"

"Aye, aye, Captain."

Everything went well and the anchor and rode were back on board.

"Tiller, easy left, now neutral. Sergeant please tell everyone to look out for rocks and sing out if any are sighted. Our lookout can teach them how to do that. Tiller, neutral, two bells Sam. Look there Sergeant at the small waterfall. That's most unusual considering a below-zero temperature is surrounding it. One bell Sam, tiller, easy left, easy right and then back to neutral. Now the engineer will give you a little more speed. Tiller, take me to that little water waterfall. Point the bow to it."

As we moved closer, Sam slowed the boat to a crawl because visibility was no more than fifty yards. I saw several places that looked promising. But first,

I had to know where that thawed ice was coming from. I ordered the lookout to use the boathook and pull us as close as possible to the little waterfall so the Sergeant could have a closer look. After he had made his survey, he reported back that he had located a small section of underground electrical conduit which had somehow become exposed.

"It is obviously energized because a little bit of ice around it is melting, but we can talk about that later. Right now we should secure the boat so that the engineer can finish his work. Is this location satisfactory Captain?"

"I believe it is if we can camouflage her to your liking, however there are no signs of tide drop so let's use 4 feet on her mooring lines."

"Will do Captain."

The Rangers began a task at which they are unusually gifted, and our engineer also began a task at which he too is unusually gifted. The Rangers completed camouflaging our boat in record time. They put snow all over the boat cover, which proved to be excellent insulation as well as perfect camouflage material, and to our advantage it was still snowing. After I was satisfied with the boat's concealed mooring, I told Sam he had my permission to shut down the engine any time he chose for as long as he wished.

He asked the Sergeant if one of his Rangers might be used to help him repair the engine. It did not surprise me when our leader answered yes.

"May I borrow him for a few hours?"

Again, yes was no surprise. However, what did surprise me was how much room in the boat it required to make engine repairs. But Sam was that kind of guy, he always carefully prepared everything before he turned the first bolt. Having settled the very important engine repair issue, the rest of us focused our attention on inspecting the boat's inside hull, cargo, and storage lockers, inch by inch. She had taken an extreme beating during the blizzard and I had to know if she were still seaworthy. After more than two hours of moving things around so that I could inspect everything, the Sergeant called time out to eat, no one refused that invitation.

Chapter 6

The weather outside was still sub-freezing with moderate snow and bone-chilling wind. Because there was no snow and no wind, it was much more comfortable inside the Ranger-built cabin, even though close to matching the outside temperature. The only problem I had was remembering not to use my hands. Everything had to be done for me. Well ... not quite everything.

During the meal break, I expressed to the Sergeant my thoughts about the exposed, energized, electrical conduit.

"I'm leaning very much toward believing that it is the power source for a lighted buoy at the entrance of the Yalu River. Tonight, if I can see a stationary light of any color in the northwest, that will be it.

"Also, the Yalu flows southeast and both light and river will be very important navigation aids for me."

"Do you think you can judge how far away it is, if we see it?"

"Yes I can Sergeant!"

"By the way, I need to take another look at your hands before you continue your inspection."

"They feel much better!" I reported.

"That is good to hear, remain where you are, I'll get the medical kit. Okay, rest your hands one at a time on this fuel tin."

As before, he slowly and carefully began unwrapping the large bandage on my right hand.

"I'm curious why you chose my right hand to examine before the left?"

Still unwrapping, he answered, "Because you are dominant right-handed. Therefore, you are likely to use your right hand more frequently than your left. Consequently, additional injuries often appear on a person's dominant side. Anyway, I did not choose your right hand. You did. Remember? I asked you to put your hands one at a time on the fuel tin, you offered your right to be attended first."

"Why the large bandage and slow removal?"

"The extra large bandages are a constant reminder that your wound is in a place that has a very high probability of further injury. Slow removal of the bandage is done so that I can smell gangrene (rotten flesh) if it is present. But be thankful because in this cold climate gangrene has little opportunity to do its mischief."

After the bandage had been completely removed, the Sergeant began a thorough examination of my right hand. "It has been approximately thirty hours since the first treatment and this hand looks good! I'll medicate and bandage it again and then have a look at lefty."

He followed a like procedure with my left hand and came to the same conclusion. "Captain if you have the same rate of progress for three full days, I predict you are out of the woods so far as infection is concerned. But I can wait because I need you if I'm to be successful on this very important mission. Just think how long it would take to replace you. However, in my opinion you are irreplaceable. So this time I'm going to give you an order, you are restricted as of this moment to limited duty only."

"What does limited duty mean?"

"It means you must not do anything without my approval!"

"Sergeant, I understand where you are coming from and I will do anything necessary to help my team complete this mission. I now know that I must absolutely follow your directions if I expect to heal enough to do that and I will."

Our leader ended lunch break by saying, "Let's get back to inspecting the boat. But remember … our captain will think and we will work."

We had so much stuff on board that what we were trying to do became very difficult. Finally, just before dark, we finished. Just minutes later, Sam asked my permission to start the repaired engine.

"Permission granted," I answered.

But he and his helper had weakened the boat's heavy-duty marine battery by overusing the two battle lanterns. It would not even begin to turn over the engine.

"Break out the hand crank," I ordered.

"It's not where it is supposed to be, Captain."

My heart almost stopped!

Thank God one of the Rangers asked, "Is this it?"

"Yes that's it. Pass it to me," Sam requested.

Things had been moved all over the already scrambled boat to create space for the engine repair. Somehow the crank had been moved from where Sam had put it. Needless to say I was filled with relief.

I looked at our engineer and said, "Wind her up Sam."

He put the hand crank on the engine's crankshaft, gave his helper instructions and looked at me with a smile. I smiled back and bobbed my head to indicate, give her a try, Sam. His helper repeatedly cranked her through many piston cycles, but she just wouldn't start.

Sam allowed the Ranger to catch his breath while he did further tinkering. With everybody's fingers crossed, he signaled the Ranger holding the hand crank to try again.

THAT TIME SHE STARTED!

Everyone in the boat had to be called down once again for being too noisy with their emotions.

Sam continued adjusting and whatever else for at least another hour, including stopping and restarting many times. When he was confident that our engine was up to his mechanical standards he reported to me that it was now completely dependable and ready for anything I wanted to do. He also asked if he could leave the engine idling through the night to charge the battery.

My answer was, "Yes, that engine is music to my ears. You may do whatever you wish concerning the engine because without it and you we all would be short timers (not long to live)."

Sam looked at me as only a close friend could and said, "Bill, we have been through a lot of dangerous things together, but I really did not expect us to make it through the blizzard because I knew that we were very close to losing our engine. I won't tell you what I found when I opened it a few hours ago but I will say that it really should not have been running at all. Something kept it running, but it wasn't me. I believe it was God. There is just no other way to explain it."

"That's exactly the way I feel, Sam. My seamanship during the storm was far beyond anything I could have done without Him. There is a reason why we are still here, so be assured that someday we will know why, of that I'm sure."

Visibility had greatly improved. Therefore, a ten-foot-high lighted buoy approximately ten miles distant from us should have been easily seen, but I couldn't locate it. Perhaps later when it's a little darker I can.

The sky was filled with enchanting stars and I marveled at them a while to pass the time. I should be able to see the light I'm looking for by now, I thought.

Then it dawned on me that a northwest bearing would be relative to my location. Since I didn't know exactly where I was, I broadened my search. In no time at all I had sighted what I believed to be my target. My reason for believing that was that there were no other lights visible, and the one I was observing was stationary.

"On-deck, Sergeant, I have something to show you."

He was by my side in five seconds. Pointing I said, "Look there. I believe I've found the entrance buoy light to the Yalu River."

"That is great Captain, have you calculated the distance?"

"I have. Ten to twelve nautical miles assuming the light is on a ten-foot tower, which is an international standard. However, it will take a little time for me to work out a safe course. But we can be reasonably sure that the exposed conduit we found energizes that light. We can also be sure that on the opposite end of the cable there is a transformer and people. Whoever laid the underwater cable had to do it with a boat, probably one with a deeper draft and larger than ours. And of course they took the shortest route because of cost and used this rock and others instead of very expensive underwater junction boxes. And since the shortest distance between two points is a straight line, that's the course I'll choose. Sergeant, I'll need you to determine from the heavens our return course."

My strategy would be to think further before sharing my complete plan with our leader, because there was a possibility that we would not return to this rock for reasons no one, at that time, could foretell. The Sergeant and I volunteered for the first security watch. That would give us plenty of time to talk about what he planned to do.

It was my opinion that we were less than thirty miles from what we were looking for, but the problem was how I could navigate on the river completely unseen. We both agreed that the staging area for the Sampans had to be up-river close to the ice barrier. However, a daylight observation was not wise because the Sampans more than likely left after dark on the first outgoing tide. The river's current plus the tide's flow would move a small muscle-powered boat at four knots or more.

"Captain, hold up there a minute, you might have hit the jackpot without realizing it! The Sampans cannot leave on an incoming tide but we can! A crack in their armor, that's what I've been looking for! Do you have any way to know how far up the river it is to the ice barrier?"

"No! But I do know that salt water mixed with fresh water has a lower freezing number, 26 degrees I think is correct."

"Can you tell about a tide schedule?"

"Well the first thing you should know is that tides ebb and flow twice in twenty-four hours. I believe I could give you six hours of darkness on the river at the right time, but tide times advance about one hour per day so I cannot promise a clock number. For example a high tide at seven a.m. today will be approximately one hour later tomorrow and et cetera."

"What's it doing now?"

"Coming in! But there's a lot more to tides than that. I'll observe and prepare a comprehensive report on local tides tomorrow."

"In the meantime," our leader vowed, "I'm going to post an around-the-clock lookout on that entrance buoy. Coxswain, tomorrow night I want you to take my Rangers and me up the Yalu River to its ice barrier because I believe that's where the staging area for the flanking maneuver is located. The time and distance I leave to the Navy."

"Yes sir! I'll be ready before noon to tell you a time I recommend for leaving."

"All right, but remember what I told you about your hands. I will check them tomorrow afternoon, and if they are like I expect them to be, we will go. But if they are not, we wait."

"I understand, Sergeant."

There is a lot to be done before we leave tomorrow, I thought. But the first thing I should do is make a record of high and low tide activity because that knowledge is the key to our success.

After observing a twelve-hour tide cycle, one ebb and one rise, I wrote "five foot" in my log book as the tide range I would use for the mission. Since we could not leave this mooring until after dark, I allowed two hours motoring time from here to the entrance light, because the course could be very hazardous, requiring slow speed and careful maneuvering. My plan was to motor to the lighted buoy and then follow the river inland on a rising tide. The Sampans should be gone, assuming they left on the outgoing a.m. tide before we were to begin our mission.

Unless there are guards along the river banks, which I doubted because of the bitter cold, we could get lucky and slip into the enemy's midst without even being seen. Then the Rangers would execute their plan. My responsibility was to take them to the fight, protect the boat, and be prepared to move out when ordered to do so.

Chapter 7

The following morning I began discussing my plan with the Sergeant, who was a very good listener, but if my plan had a flaw he would instantly know it. When I recommended leaving the entrance buoy at approximately nine p.m. on an incoming tide, he challenged me to support my position. I did so with everything in my log book.

"Excellent job coxswain! We will leave from the lighted buoy at nine p.m. By the way, I very much like your plan to move our boat to the buoy at almost low tide so that dangerous rocks will be exposed and therefore avoided."

"Thank you Sergeant, and now for the icing on the cake. We are going in on an incoming tide, which will give us six hours before any watercraft, except one with a motor, can cause us a problem."

"Good," he laughed, "but I will complete my business in less time than that."

As promised, the Sergeant began his evaluation of my injured hands. He requested that the tiller be brought to him, then told me that my hands were steadily improving, but he was going to bandage them a little differently. Instead of a useless ball, he made the new hand wrappings like cups that fit nicely around the tiller, and cautioned me to be very careful. Then he began wrapping the tiller with gauze and reminded me that this time I could use my eyes, not my feelings from the tiller. I was very happy that I was well enough to go on the mission.

We all ate and drank plenty, as instructed by our leader, after which everyone continued preparing the boat and themselves for the mission. The Rangers waited until the last minute before removing the boat cover, so they could have better firing opportunities from the boat with their weapons. Discontinuing our camouflage gave me some concern because it was not yet dark. However, we were in one of the most uninhabited places on Earth and it was my opinion that if there were humans about they would be trying to keep from freezing to death instead of being out in freezing weather looking for spies. For that very same reason I believed we would catch our enemies at the river's ice barrier off guard. Unless we froze to death before we got there!

The Sergeant and I had taken noon sights with my sextant and everyone set their watch accordingly, twelve noon. As long as we could see our watch, we had the correct time with plus or minus five minutes meridian accuracy. Our leader made sure that each man still had a cyanide pill by having him show it. But I had a problem because of my bandages. The Sergeant soon took care of that issue by taking my pill from my clothing and putting it without the wrapper in a dry part of my bandage that I could easily reach with my mouth.

We were all at a level of excitement which was difficult to manage. I had planned to leave our mooring at seven p.m. After waiting for what seemed to be an eternity it had finally gotten to six p.m. One more to go, I thought. Our leader called us together at six-thirty p.m. and invited us to join hands for prayer. We all did so and he prayed that God would be with us and give us victory over the infidels who had murdered His servants.

The team was on fire, ready to go. The Sergeant was a master at getting his people charged up. Therefore, I would wait no longer.

"Are you ready to go to war Sam?"

"I'm ready Captain! Start your new engine."

He smiled and answered "Aye, aye Captain. Stand by all lines fore and aft. Sam when you're ready, give me three bells slow."

As the boat started to move out of her mooring I shouted "Take in all lines fore and aft."

I brought her about and hesitated until I could chat with the bowman, essentially to tell him that it was his boat until we reached the lighted buoy. "Sing out your warning to Sam and he will do the same to me."

Thus began a very risky passage to the Yalu River entrance, which without rocks could have been done in one hour. To this day it still amazes me to think that I raced across those rocks at full speed without putting a scratch on my boat. Don't tell me that God doesn't look after His children. The two hours that I had allowed to cross the ten miles might prove to be insufficient, I thought, if the first hour was an indication of what the last hour would be. But I knew I must not be impatient with my crew, because no one could do it better than they. However, it did seem that we had begun to move a little faster in the past few minutes.

Sam and our lookout had been communicating all along the way while moving the boat from one rock to another, but apparently it did not require my involvement. The fact that we were steadily moving at any speed was a blessing. It sure beats being grounded, I thought.

The closer we got to the buoy the fewer rocks there were. In fact, we could expect the wide river channel to be completely rock-free. We reached the buoy light's coverage area safe and sound at nine-twenty p.m., which pleased me. Pointing to the light, I advised Sam to get us away from it fast, then throttle down.

That brief speed run was a good test of my steering ability while wearing the cupped bandages. So far so good, I thought. However, this late in the game I would not tell the Sergeant even if I did have pain.

Our boat was in a good position to begin our trip upriver in search of the Sampan staging area for the communist flanking maneuver. But before starting, I thought it would be good to advise everyone about what I planned to do in terms that would assure their complete understanding. Because even a small misunderstanding could botch the entire mission and I did not want that to come home to my door.

"Sam, let's 'heave to' (stop) for a minute."

I brought the bow around to face the current so it would be easier for the engineer to hold our position.

"On-deck Sergeant, for a quick strategy session. Please move everyone aft so they can hear my voice. Rangers, we are presently experiencing a mid-tide current which is the swiftest hour of any tide. Assisted by a fast current, I can travel fifteen miles per hour. That means we could be looking at our target in one hour or less. The reason I believe that it's one hour or less away is because sea water has a large salt content and freezes about six degrees below that of fresh water. But it is highly unlikely that sea water will influence fresh river water more than ten miles inland so that is where we will probably find the river's ice barrier and the staging area we are looking for. Unless there is something unknown to me now, I can easily be at the enemy staging area in an hour.

"And another thing. Sneaking in might telegraph that we ARE an enemy, but going in fast without concern for engine noise might send the opposite message. Fast or slow – what will it be, Sergeant?"

The Rangers mumbled while huddling a minute or two before the Sergeant answered back, "FAST!"

"All right, fast it will be. But I will need everyone's eyes to watch for anything adrift on the river that could create a problem if I collide with it. My primary concern will be ice that has broken free from the ice barrier and is on its way to the sea. Our leader has instructed me to land on the North Korean side of the river which I will do. But be assured that the Navy will not leave, no matter what happens, until our team leader has ordered us to do it. We have never left a Ranger stranded and this mission will not be the first to do it. I'll come about now and prepare to start up the river after everyone has returned to their assigned place to defend our boat. It's my intention to take you Rangers on a fast boat ride up the Yalu River that will not only surprise our enemies, it will shock them! We will attack so fast they won't be able to fire a shot. One bell full ahead, Sam!"

Chapter 8

As our speed increased, I imagined our boat being a raging bull cutting through the water in pursuit of an enemy who is waving a red flag. After only a few minutes I dodged an easy-to-see block of snow-covered ice which was pointed out by my lookout. That one was easy to maneuver around with no slowdown. During the high-speed ten mile run there would be many more, but I didn't reduce speed to dodge even one. Thankfully, the newly repaired engine at full throttle was running unusually quiet, which improved our communication. But nobody was talking.

I could not think of anything except what was immediately ahead of my speeding boat, second by second. The Yalu at its delta is about three to four miles wide. As I sped inland that distance narrowed to about one mile or less. Therefore dodging the ice became increasingly more difficult but I still did not slow down. The center of the river was the best place to run fast because drifting things tend to migrate to the edge in moving water.

Smoke from burning wood caught my attention. "We are ten minutes out," I shouted to everyone.

My eagle-eyed lookout shouted back, "NO, five minutes."

"Did you hear that Sam?"

"Yes, I'm going for five!"

"Okay, the boat must be landed on her starboard side with her bow clear to run down river for a fast getaway." Sam pulled back on the throttle and put her in reverse full power to slow her, so I could turn port and find a landing. In freezing darkness except for a little bit of moonlight, I shouted, "Landing in sight" and steered for an uncertain landing at a little wooden pier, which I almost destroyed with a hard landing.

Sam held her in place as best he could by manipulating the engine, while the lookout and I held her with our hands against what was left of the little pier. After I shouted, "Starboard landing secure," the Rangers jumped onto the flimsy pier, running. They were gone in a flash. The lookout and I continued holding our boat against the pier, which was very inadequate for docking a U.S. Navy assault craft, while Sam used the engine to take the strain off of our energy expended fighting the incoming tidal flow.

In only seconds after the Rangers left, we heard grenade explosions and small arms fire as if it were the Fourth of July. There were ever-increasing fires all over the place and absolutely no let-up of gunfire and explosions for five minutes or more. From the lightning-fast beginning, I heard desperate people who were likely fleeing for their lives, screaming, but apparently to no avail because gunfire and frantic human voices seemed to calm down almost simultaneously.

The Rangers, with their pathway and our boat lighted by fires they had started, came running to the boat carrying several boxes. They immediately manned their onboard defense positions, because light, in our business, is an open invitation to bullets.

The Sergeant, with one big breath, shouted, "Everyone is back on the boat, get us out of here fast Captain!"

"Will do Sergeant!"

While I was saying those words, Sam was already advancing the boat's throttle to full ahead. He was no doubt the most important member of our team - he could read minds! A minute or two after we had begun our race to the Sea of Japan, I overheard the Sergeant say, "Seven minutes, people, and not a shot was fired at us."

Everyone did their job perfectly on this mission and they should be very proud. However, I had much trouble being proud of killing anything, especially people. As I have written earlier, I'm not a true warrior! Therefore, I grieve easily and deeply when people suffer an untimely death.

Coming upriver I was firmly focused on coursing my boat on its middle and did not turn away to search for Sampans. But going down river on the middle, it was easy to see by moonlight many Sampans tied along both banks of the river (Russia and North Korea).

"Look at all of those Sampans Sergeant, that proves that we have hit the jackpot!" Our team enthusiastically cheered our leader because his remarkable leadership had made it happen. We all had gone through much painful suffering and unbelievable danger to accomplish what we had done, but nothing could equal the joy of success that we were feeling at that moment.

In spite of motoring full speed against a two-knot tide, I saw the entrance buoy light in little over an hour. "What now Sergeant?" I asked.

"Go out of the buoy's lighted area as far as you feel it's safe because they could be waiting for us to appear within its beam. You can bet the Russians have already put helicopters in the air with spotlights to look for us. They will soon discover that we came up the river, which they will search thoroughly, and be prepared to use fifty caliber machine guns if they find us."

We sped safely past the buoy, and the Sergeant gave me a star to steer by within minutes.

"Southeast for now, Captain."

"Understood. Southeast, Sergeant, I'm on the star."

Our leader continued, "I want this boat to be radar-proof – because the Russians will really be after us, now that we have proven that their coastal defense is not impregnable. First, remember that no metal will be left uncovered. Second, the boat cover must be put back as it was because it has proven to be the best radar protection that we have. Let's get it done, men!

"Captain, so that you will know what I'm doing, I'm trying to put as much distance as possible between the Russians and us, so steer a straight course until I change it."

"Will do Sergeant!"

Before dawn the Rangers had finished their restoration of the shelter and they had a warm cabin again. Notice that I wrote "they" because I did not help.

The Sergeant approached me and asked my opinion about our situation.

"Well Sergeant, my guess is that we are now approximately thirty miles off the east coast of North Korea, but landing anywhere up here would be suicide because sooner or later they will find us. Russia and North Korea are militarily embarrassed that we were able to pull off such a perfect mission that left them with egg on their face, and that does not make an Oriental happy. Notwithstanding the fact that hundreds who were asleep at the wheel have already without a doubt been summarily shot, their determination to get their hands on us grows by the minute.

"Consequently, North Korea and Russia will make an all-out effort to find us. But if that doesn't happen, they will try to cover up their vulnerability by claiming the UN bombed Russian territory. So any way we look at it we are in a big, dangerous mess. I believe we are doing exactly what they will predict, escaping south where the big search awaits us. So let's outfox them again and go east to northern Japan. No other coxswain in the world would dare attempt that winter crossing, especially in an open-hull whale boat. But that strategy offers a much needed benefit, our enemy is less likely to risk aircraft looking for us in that cold no man's land."

"How far is it?"

"Four hundred miles more or less twenty."

"Do we have enough fuel?"

"Sam, do we have enough fuel to reach Japan?"

"How many miles?"

"Four hundred," I answered.

"Wait one! Yes we do Captain if I use sixty percent throttle."

"How many knots will that give us?"

"Seven to eight."

"Thanks, I'll be back in touch, but in the meantime, run full throttle."

"How long will it take Captain?"

"I'm sure you can tell that I've already done some calculating! The answer is one and a half days if we motor around the clock. Our greatest risk will be today (6 a.m. to 6 p.m.). If we survive that time period, I feel confident about our chances to reach Japan safely. The bloody communist will be trying to find us for weeks along the entire southern coastal route. Therefore, if we want to keep living we have no other choice but go to Japan.

"Sergeant, I need your permission to change our course and steer due east to Japan."

"Permission granted, Captain! Is there anything we can do to help with the dangerous daylight hours that are only minutes away?"

"Well yes Sergeant, there are several things that can help us survive. First, since our Navy people are all on watch, I want the Rangers to make a Japanese flag for us the same size as the American flag we have on board."

"Where is our flag?"

"In the forward storage locker on the starboard side. An important reminder just occurred to me, Sergeant. We must not send or receive any wireless messages today, because that could disclose our position."

"Aye, aye Captain."

"Perhaps the most important thing now is to make our boat appear, as much as possible, like a Japanese fishing craft. No more than two people visible at any time. It is also very important to splice all of the earlier cut DC wires back together, so that if I need to imitate a small commercial vessel, I can. Please connect the engine hour meter first, it will display the total time our engine has been running. And then the RPM meter, so that our engineer will be able to monitor our fuel consumption and the optimum speed we should maintain. Next, set the engine hour meter on zero and use your watch to record the time it began. After doing that, I can give you (our leader) an approximate location for your wireless report at any time. Finally, I don't believe I can survive much longer at the helm in this bitter cold, so I have an idea that will help solve that. After the other things I've talked about have been done, I'll share my idea with you."

Chapter 9

Dawn was peeping over the eastern horizon. "Hope I can see the same beautiful scene tomorrow," I whispered. "Sam, keep us full ahead throughout the day, if the weather permits, and we will talk further tonight about the fuel situation, if we are still alive."

The seas appeared to be making one to two-foot waves, therefore we were leaving Russia behind at ten to twelve knots (12 mph). And that was exactly what we had to do – get out of the communist search area as quickly as possible if we were to have any chance of saving our lives.

Daylight duty on our boat was so unusual that we all had trouble adjusting to sunlight. Thoughts of being in the comfortable cabin, with my back against the warm engine housing, made me drowsy, but I dared not sleep. So to stay awake, I began talking with Sam.

"If we are attacked by helicopters a zigzag course to avoid their bullets would be all I could do. But I'm sure the Rangers have their own defense plan if needed. But reality is we would not have even a remote chance against a helicopter. However if we make it through the next twelve hours without being discovered I believe we will be safe. Because our pursuers probably would not go beyond one hundred miles east of Vladivostok, which is for us ten hours from the lighted entrance buoy, or ten a.m. clock time. But I cannot relax until darkness once more covers us because that's the only thing we can really count on. But if we don't make it, Sam, I'll see you on the other side."

"Don't forget to pray!"

"That I will do, Sam, that I will surely do and you do the same."

I decided the time had come to discuss with the Sergeant my plan to move into the shelter, an idea that would surely keep me from freezing again and consequently becoming useless to the team.

"On deck Sergeant, I need to chat with you."

"What is it, Captain?"

"Since I can see the boat compass and even your field compass in daylight, I can steer from under the shelter and get out of this cold wind now, instead of waiting until later. Here is my plan. Find a small gauge rope about twenty or more feet long, and bring it to me so I can demonstrate my idea."

He quickly returned with a rope similar to what had been requested. After examining it, I reasoned it would do the job nicely. But since my hands were bandaged, I used his.

"Put the middle of this rope over the forward end of the tiller. Tie a clove hitch in it and pull it real tight. Good! Now pass the two ends through the railing ring bolts (pad eyes), one on the left and the other on the right, which will form two perfect right angles. Take the two bitter ends and stand by the engine with one line in each hand. Gently pull the tiller left and right for a test."

"Very good! Now I don't have to worry about my captain freezing to death."

"Thanks but I'm not quite finished. I'll need a place to sit, and two small cleats to secure my end of the tiller control lines (ropes) during heavy weather.

"Now the last thing is to bandage the clove hitch tightly in place to insure that it does not drift aft on the tiller."

In thirty minutes the entire process was completed. All that was left was for me to move into the warm cabin and continue steering east. Since the seas were mild, steering with the newly created apparatus was easy. Just getting out of the arctic wind was a blessing in itself, but throw in a little heat from our engine and that really began to warm my bones. The Rangers had removed, from the engine housing, not just the inspection port but the entire top, which released much more heat. The rising sun was barely above the horizon and it appeared that we were in for a cold but beautiful day. Which was not good for us because there was no place to hide, and the cruelest people on Earth were desperately searching for us.

We had been running at full speed for about seven hours, through darkness into the morning light. We were near seventy miles east of Vladivostok, in plain sight within a radius of ten miles. The only hope we had was that our wooden boat would continue to shield us from Russian radar. That's why, from the beginning, a whale boat was chosen for the mission. My guess was that Russia's search for us would be limited to one hundred miles south of Vladivostok and fifty miles east of that city on the Sea of Japan. With their national pride at stake, it's easy to believe the searchers would be very thorough, since their lives might be the price they would pay for not capturing us, dead or alive.

If they failed, Russia would likely begin a verbal storm of propaganda, claiming the UN had bombed targets within their so-called impregnable coastal defenses, and had also attacked, with ground troops, a river docking facility in Russia which was being used to transport food supplies to starving Chinese people. Their purpose would be to cover up the truth and make the UN appear to be the bad guys. But even if the truth had been known, Russia would still deny our success, because that action by five U.S. Army Rangers and three U.S. Navy sailors had destroyed their boast of impregnable sea coast defenses, and had further proved that Russia and China were very involved in the Korean War.

The U.S. Army Rangers and three U.S. Navy sailors who had made that attack were the bravest men I had ever heard about, read about, thought about, or which had come to me through any manner of communication. My admiration of the bravery of those men was based upon an eyewitness account – mine. Even after realizing that this day could very well be our last one on Earth, I was still very glad that I had been chosen to be the Rangers' boat captain. I repeat, where does our county get such men?

"How is this, Captain?" the Sergeant asked while holding a rather good imitation of a weather-worn Japanese flag before me.

"Perfect, and in good time! Deploy it right away in the flagstaff holder aft of the cockpit, and then come back for a chat."

"It's done, Captain, we are flying a Japanese flag," he soon reported.

"Good! Now I want you to pick a man, no volunteers, who can best mimic a Japanese fisherman. After preparing his costume, which should be ragtag, put your chosen man in the cockpit, with his arm lightly resting upon the tiller as if he were steering the boat."

"Okay Captain. I'll have fun with this one."

It was close to an hour before our leader returned and presented his made-up fisherman to me.

"He will do very well," I predicted.

"Sergeant, tell him to give a friendly, fearless wave to any aircraft that shows an interest in us, but if that wave draws it close enough, I'm sure you will know what to do."

"We've already prepared for that, Captain!"

"Why am I not surprised to hear you say that, Sergeant? Our fisherman needs an identity, let's make him a troller. Use the oars, boathook, or whatever to simulate trolling poles. If we appear to have attracted someone's attention we must reduce speed immediately, because we are moving too fast for a trolling vessel. Also, speed will make significant wake and that's another thing we must be careful about. We must not appear to be running away."

For those readers who are not familiar with boats, wake is the white foam created by a fast-spinning propeller.

"Sergeant, it is ten a.m. and we are now one hundred miles east of the lighted buoy, however, I will not feel safe until dark, eight hours from now. Then we will set up an at-sea watch routine, and I want it to be two hours on duty and two hours off. But for now every man is considered to be at a battle station, sweeping the sky and sea constantly for those who want to kill us. Have them report anything they see to you because I would like for you to handle those things for me."

"Aye, aye Captain."

"It is my intention to continue full speed ahead as long as possible."

It seemed to be a good time to test how much trust I could put in my irregular steering apparatus. I advised Sam that I would be making moderate course changes for a while. Very soon, I realized that using the makeshift steering apparatus caused the compass to delay answering the tiller by perhaps five seconds. I could get into big trouble with that problem in a stormy sea, where split-second timing is a must, especially in an overloaded boat. No matter what I tried, the necessary feel of tiller reaction could not be found. On seas with four-foot or more waves, there was no substitute for my being at the tiller to make immediate corrections. With anything less, the boat would be lost at sea and we would freeze to death in water that, at that moment, looked so beautiful and full of life.

"On deck, Sergeant, I have discovered a potential problem."

"What is it Captain?"

"A delayed compass response to the rudder, which cannot be tolerated in a stormy sea. That means that I must be at the tiller in all weather worse than now. Since it is always necessary in our overloaded boat to run before an angry sea, I cannot use the compass. Chances are the weather will get worse during the night and since we are so close to safety, I want the Rangers to make a survival suit for me ("Captain Caution") so that I can make it through the coming night at the helm (tiller) without freezing to death. I have decided that tomorrow we will be close enough to Japan to call for air cover and rescue."

"Captain, since the seas are now calm enough for someone else to steer, I need your body to tailor something for you."

They made a warm bed for me on one of the forward seat boxes, and I was so exhausted that I was asleep in seconds. When the Sergeant woke me four hours later, I noticed our lookout manning the tiller. I smiled at him and he responded in kind. The Sergeant had obviously begun the at-sea watch schedule. It was just past four p.m.

Our leader informed me that everything had gone well while I slept. The Rangers began feeding me warm water (melted snow), K-rations and candy bars. The reason they do not drink coffee on a mission is that its aroma is long-lasting and far-reaching.

"How many hours from the buoy, Sam?"

"Seventeen hours and six minutes on the engine time meter, which includes estimated time from the buoy until its hook-up.

"How about our fuel?"

"I filled the tank (20 gal) while you were sleeping and there are one hundred gallons left."

"I plan to see the mountains of northern Japan by noon tomorrow, if we run hot until then."

"Not recommended, Captain! I can run full throttle on what's in the tank now, which will be six hours, but after that time I must reduce speed to three-quarters throttle or less."

"Thanks, Sam, you have my permission to manage the engine priorities as you see the need. Another thing Sam, the Sergeant and I have agreed that it is best to turn on all navigation lights during night hours, because Japanese fishermen with no lights on their boats are abundant from here to their coast. It might be wise to energize them all now to be sure they are operating properly."

"Sergeant, please remove the Japanese flag and replace it with our own at first light."

He answered, "Will do, Captain" and continued, "It's time now for us to put you into your survival suit, which must be done in layers."

"All right, mates but maybe I should answer the call of nature before being put into that suit for eight to ten hours."

"Good idea, Captain!"

While I was being dressed, the Sergeant informed me that he would check my status in the cockpit often. Everything possible had been put on my body, excepting my hands, to keep me warm. Then the Sergeant satisfied my curiosity and began to explain how that special need would be handled.

"I will remove both bandages and determine whether or not I'm going to allow you to do this! Which one should I examine first?"

I held up my right hand, which had been hurting since holding the boat against the little pier during the attack, but I held back from telling the Sergeant about it because it happened just as he had predicted. Even though at that time I was in considerable pain, I would not fail my teammates at such a critical time. We had to run for our lives as quickly as possible and had no time to care for an injured man, especially the captain. After removing the bandage, the first words he spoke were "DAMN! Why didn't you tell me about this?"

"Is it bad?"

Well it certainly is not good! Give him morphine, I've got to work on this a while. Now this is going to hurt but it must be done. But before I begin, I must tell you that I understand why you acted as you did, and if you had not done so, none of us here would be alive. Your amazing seamanship has brought us within hours of safety but I will not allow you to suffer like this another second! Here, take this pill. As I have previously told you, we need your brain, your experience and not your dead body."

"I understand that Sergeant, but I need help to get my boat through this last night, it's very important to me."

He was correct, it did hurt, big time!

The lookout who was steering called out, "I can't see the compass and I need a star to steer by."

The Sergeant left the shelter, quickly fulfilled the tiller watch request, and returned to his patient. After cleaning, medicating and bandaging my right hand he began the same procedure on my left hand, which did not hurt at all. In a short time he told me there were no gangrene indicators on my left hand but I was sternly warned to use it only with extreme caution. Outside the shelter it had gotten dark and colder and I felt very uneasy about not being in the cockpit. I asked the Rangers to complete dressing me so if it became necessary, I could at least stick my head outside.

"That will be my call not yours!" our leader made emphatically clear, and followed with, "Finish dressing him."

"Thanks Sergeant! The boat is surfing fast, I can feel it and I'll wager she's making hull speed (as fast as she can go). I must tell whoever is at the tiller not to over-control!"

"Finish what you're doing, I'll tell him!" our leader assured.

"Sam, if we keep this up we will see Japan at daybreak. While she's planing like this she is using less fuel so keep her at full throttle!"

"Aye, aye Captain!"

"Sergeant, we must prepare for a large number of unlighted fishing boats, which are a big navigation hazard to say the least. Japan's northwest coast is famous for its excellent cold water fishing, that's why there are so many boats up here."

Better boats than bullets, I thought.

"I want all navigation lights on so we can be seen and then they will clear our way. Please relieve our best lookout from the tiller and put him on the bow. You and I will take the tiller – my brain, your brawn and your permission."

"Aye, aye Captain. But remember, do not use your right hand!"

"I promise! Sam do you need help refueling?"

"While she's running this smooth I can handle it with no help."

"Right, sing out loud and clear if you need anything because I intend to take us on a fast boat ride to Japan!"

The Sergeant and I quickly agreed on a tiller technique that served us quite well. At our current speed, loss of control could have easily swamped us, so from the beginning we made sure that could not happen. How did we do that? We double-wrapped the apparatus lines through the railing eye bolts and they acted as preventers which would keep the rudder under control. After six to eight hours, we began to hear warnings from the lookouts but the one I carefully tuned in was Hawkeye. He was the next thing to having a spotlight. His warnings were always clear and timely – unidentified object dead ahead five hundred yards, easy port – and so on.

More and more navigation warnings were being reported. Consequently, I called for slow speed.

"Give me a depth reading, Sam!"

"One hundred twenty-five foot, Captain."

I ordered two bells (neutral).

Chapter 10

"Sergeant, we are now in Japan, free of any threat from Russia! No more hiding and running at least for a while. May I have your permission to turn off our navigation lights, turn on our anchor light, and drop our anchor?"

"Permission granted, Captain."

"Since you have ordered only light duty for me I will obey, therefore I must step aside and leave the anchoring chore for your pleasure."

That performance broadened his smile and he asked, "How much rope (rode) Captain?"

"All that we have, two hundred and fifty foot."

"After that has been done set the anchor watch detail and let all off-duty personnel take a long well earned break."

"Aye, aye Captain."

It was a very cold and snowy dawn. Our leader ordered me to warm up by the engine, which was sure hot. That was fine with me because it would give Sam and me some private time to talk. As I walked under the shelter Sam was inspecting his idling engine.

He looked up, smiled and said, "We beat the odds again, Bill! But I'll tell you now, I did not believe this time it would happen. I can't stop thanking God for delivering us from certain death."

"I feel the same way, Sam. Do you realize I almost froze to death three times on this mission?"

"Speaking of freezing, how is our fuel situation?"

"Three tins (30 gallons) plus what's in the tank."

"Is that enough to keep us warm?"

"Oh sure, I can idle two days on that much fuel."

"Well, that might be necessary because I do not intend to get under way again with this exhausted crew. And I won't know until our leader makes wireless contact when we will be relieved."

Everyone came under the shelter brushing off snow and crowding around the hot engine. The Sergeant was back in his element and began getting things organized. First, he asked for our location and that question was directed to me.

"How many hours from the buoy are on the engine meter, Sam?"

"Thirty-three with no variables because we were running full throttle all the way. Sergeant, it's too overcast to use the sextant for a latitude fix, so I must use dead reckoning instead.

"We anchored in Japan at five-thirty a.m. this date, among a large number of Japanese fishing vessels. We are three hundred and ninety-six miles (estimated) east-southeast from Vladivostok, Russia. My anchorage depth is one hundred twenty-five foot, my vessel and crew are at this time not seaworthy, and I cannot safely move from this location. Colors (U.S. flag) were displayed immediately after anchoring."

"Thanks Captain. It looks as if rough weather is coming in so I hope I will be able to use the wireless. But while I'm trying to do that, everyone must make a supreme effort to keep cold air out of our shelter."

One of the Rangers had been using the built-in battery charging apparatus on the wireless for a good while, so the Sergeant would have no problem with power. My right hand had begun to hurt more than ever but I knew that our leader would make that an important part of his report, so I remained silent. There was an unexpected tapping on the hull. The reader would never guess what it was – a five-gallon metal container of hot tea and an even larger metal container of recently cooked fish being delivered by a Japanese sailor! His large fishing boat was anchored very close to ours, but not dangerously so. He had used a small tender (boat) to visit us. Verbal communication between us was not necessary because a book could have been written from the facial expressions of the giver and receiver. Thankful Americans and happy Japanese waved vigorously and hurried to get out of the bitter cold.

Since I was restricted from lifting, I just watched, but there was nothing wrong with my ability to smell the cooked food which was about to put new life into our weakened bodies. Now believe this or not, both metal containers fit perfectly on top of the diesel engine. What are the odds against that! And there was a cloth bag tied to the spout on the tea container – it held eight small tea cups. Why not seven ... they could not have seen Sam! And how did those kind Japanese people know that we did not have anything on board our boat with which to drink tea? We drink from our canteens! And was it a coincidence that we had motored four hundred miles and anchored our boat ten feet behind thoughtful Japanese people who would feed eight emaciated sailors the most nutritious food on the planet – fish? I don't think so!

We decided to feed the anchor watch first. I offered to relieve him but everyone refused to hear anything further about that. The Sergeant could not do it because he was busy with the wireless. Sam could not do it because he could not leave the engine. The remaining four drew straws to pick the lucky one who would relieve the anchor watch. However, drawing straws was not the issue, it was about how deeply those young warriors were concerned about the welfare of every member of our team.

The relieved watchman came into the shelter almost frozen to death. I was shocked and ordered, "No more outside anchor watch, get his replacement back in the shelter now! I'll figure out a way to observe our anchor from the inside and stay warm. But the first thing done will be to double-cleat the rode."

The boat cover was made from very heavy gauge waterproof canvas material and the Rangers had made it airtight, and let's not forget several inches of snow which further insulated everything. I asked my lookout to make a fast trip to the bow and double-cleat the anchor rode while the rest of us were wondering how to eat stewed fish without utensils. We had already begun drinking the wonderful tea using the small cups, but that kept us busy with refills. Finally, after experimenting with everything we could think of, we decided to use the little cups as spoons, assisted by our hands. I couldn't quite manage, so others helped me eat. There was just no other way, and we were very impatient diners who were ready to get on with the meal. Our leader had completed his wireless report and joined us for a party.

"I'll tell you everything after we eat. But for now, I wish for everyone to join me in prayer. No hands please, they are too greasy."

We were all anxiously waiting for the dinner blessing to end, although it was a good one, so we could begin eating. I suspected the Sergeant was holding back some exciting information until we all ate and drank our fill, even though we had a crude way of doing it. The fish gumbo, or whatever its name, was outstanding. The proof was that there was nothing left in the pot. Eating a delicious meal after having been fed K-rations for so long made me nauseous, but it soon passed.

After the meal, the Sergeant began sharing with us what he had learned from the wireless. "Ranger Intelligence Headquarters (RIH) thought we had been either captured or killed because they had been unable to contact us for days. We had created an international debate after our attack at the Sampan staging area. The UN desperately needed more proof that Russia and China have been from the beginning supporting North Korea's invasion of South Korea. Well I have that proof in those (pointing) boxes. RIH was almost sure that we had made the attack because they knew we were in that area for that purpose, however since they could not make contact, they assumed that we had been captured, or worse, killed. After I told them some specifics about the attack and the boxes, they ordered me to keep my people quite out of sight and close to the wireless. And to be prepared to kill anyone who attempts to come aboard our boat without the wireless password!"

"'We are on our way Sergeant!' they promised. Captain, you will be first to be evacuated!"

"How long will that be?"

"Soon, because our medical people were very concerned when I explained our captain's injury. All right Rangers, lock and load and get ready for a fight.

My number one order is that no one comes aboard this boat and remains alive without a password!

"Almost forgot, Captain, RIH found us in minutes using your dead reckoning navigation report. We are in a location which has a large fish processing factory. That explains the large fishing fleet."

"Sergeant, with the weather getting worse by the minute I seriously doubt if anyone will go fishing today! But that's good for us because we could be blocking a channel, and we can't move our boat."

A small red light appeared on the wireless and the Sergeant jumped for it, put on the headphones and checked in.

"Gentlemen, there has been a change of plans," our leader reported, and continued, "Two helicopters will land near here in one hour to evacuate us all. Japanese police will be at our boat in a few minutes. They have our password and will tow our boat with us hidden on board to the fish factory dock where we will go ashore and rush to the helicopters. Everything will be left on the boat except the intelligence boxes. I repeat, everything! The boat will be recovered later and your personal belongings will be returned to you."

"The police are here Sergeant!" I whispered.

Our leader looked outside and heard "Nando" to which he responded "Nando."

"Must move boat to factory, no can get anchor, we cut."

They did that, and towed our boat to a place at the factory where three Rangers were on the dock waiting. They gave the password, came aboard and took possession of the intelligence boxes. We went ashore with them and ran, which wasn't easy, to board the medical air rescue crafts. Mine was airborne before I was completely strapped in. Immediately my hand was lifted up and extended by some kind of medical device and an Army doctor began removing the right-hand bandage while an Army nurse was administering shots that he ordered and the Sergeant was answering questions.

"I can smell gangrene," the nurse said.

"So can I," the doctor replied. "Sergeant, how long has it been since it was frozen?"

"Four and a half days, sir."

"Tell me the complete treatment history."

After hearing what he had requested, the doctor made an in-flight diagnosis and recommended by radio immediate hospitalization for me. I was happy to learn I was going to Shiga Heights with the Rangers. It is in northern Japan but not as far north as we were, and thank God for that. We stopped somewhere along the way to refuel and then moved along quickly to our destination, which was accomplished late in the afternoon. In no time at all I was looking up at blinding lights in an operating room. It was obvious that I had been expected because people were immediately all over me doing this and that while taking off my filthy clothes. It seemed that every available hand was rushing to prep my body so the medical people could take over and do their thing. Just as the mask was put on my face, I overheard a conversation between those around the operating table debating whether or not my arm could be saved. When I woke up the following morning, five Rangers and two sailors circled my bed sporting big smiles.

"Well?"

"It's still there Captain!"

The doctor who had performed my surgery walked into my room soon after I awoke and pleasantly said, "Good morning all," and began reading my medical chart.

"Young man I see that you have been with us before!"

"Yes sir! Twice."

"Well, how are you feeling this beautiful morning?"

"So far so good sir," I answered.

"You have probably noticed that I left your right arm in place against the advice of all my colleagues. My assistants had already laid out the tools to remove it but a more powerful voice told me not to do it. And I always obey my coach. After getting deep into your problem, I realized what a real professional job your Sergeant had done. Without his skill under such unbelievable conditions you would be dead by now!

"Sergeant, since you have saved his life, I'll assign this case to you. I believe he would like that and remember, I'm available anytime! Report to me daily or when necessary and everyone have a God-blessed day," he said while walking away.

"Thank you, Major!" I said.

"And thank you, young man, for the service you have rendered to our country."

There is no way I can lose my arm now, I thought, because that man has a terrific coach! Two young female Japanese nurse assistants came into my room and reported to the Sergeant. After a brief talk he sent them to the dining room to get my breakfast. While waiting for them to return, all of my teammates gathered around my bed with the Sergeant acting as their spokesman.

"Captain, we have been together for six months fighting a bitter war. During that time, you and your crew risked your lives time after time to save our lives and boasted that the Navy had never left a Ranger stranded in North Korea with no way to get home. And now the Rangers are here to say that we will not abandon you in your time of need. Our goal for you is to walk out of this hospital with both arms swinging. The methods we use may not at times please you, but they will certainly be done for your best interest. For example, everything you do will be closely supervised twenty-four hours a day for ten days. Both nursing assistants will remain in your room for that same period of time. I will not go over your expected behavior concerning these ladies because you have heard it all before, but remember they are very important to your recovery. Also, there will be a Ranger posted in your room every minute of the day to insure that you do not, for security reasons, talk to anyone."

The ladies returned with my breakfast and I was eager to begin. However, my fragile right hand was temporarily secured so that I could not accidentally use it during my meal. The Japanese ladies had prepared my food so that I could with only a little help feed myself with my fifty-percent usable left hand. With all beverages, I required help. I felt as if the medications used to control pain during my surgery were almost gone away because my mind had become more and more clear. The reality of my current situation became clear – I am no longer freezing and fighting every minute to keep my team and myself alive, I am warm, I am safe and well cared for and I still have my right arm. What more could I ask? Thank you God!

On our mission to the frozen north, I came dangerously close to freezing to death three times, a horror which I will never forget.

After my meal, the Sergeant continued. "We will be sealed off from everyone until Russia's propaganda war at the UN is either won or lost. But if they lose, just imagine how much more determined they will be to get their hands on us. We have enough evidence but it will take time to prepare it for use in a worldwide forum like the UN At some future time there will be an attempt to separate and question us, however, I will resist separation as long as possible. But remember this, it is highly possible that we could be moved to another location before being interviewed. That could really throw the hunters off track and that I would like! And don't you forget that the Russians are still searching for us and will continue doing it for a long time. Russia is famous for that. The loss of their worldwide invincible image, thanks to a small boat and eight men, will not be soon forgotten by the world's military giants. The Russian Bear will now have much less clout at the UN."

The Sergeant with a laugh said, "By now they probably know our mothers' maiden names and where our grandmothers are buried. Because there is a good chance that our boat, which was left at the factory, has been photographed inside and out, stripped of all useful information, and the results quickly sent to Russia by Japanese communists. One thing more, Captain, our team is now the only occupant on this "closed for repairs" wing of this hospital. For our safety, we don't exist, now or in the past. That means no records!

"Now I'll get back to your medical case, Captain. I'm sure you are curious about the large bandage on your right hand, which feels as if it were in a cage. Well, it is in a cage because any injury, no matter how slight, could cause you to lose it. Therefore, I plan to secure your lower right arm across your belt line so you will have much less opportunity to injure it while sleeping. You must sleep only on your back. The nurse assistants will be posted at your bedside while you sleep to make sure that you do not roll over or get up without their assistance. Whatever independence you harbor or privacy you prefer, forget it for ten days or you will lose that right arm! Be patient with us because we will be very dedicated to our purpose."

Chapter 11

"As you well know from your previous visits here, you were permitted to order food or drink anytime you wished. However at this time I must deny that privilege because in your weakened condition you would not survive an additional infection. Consequently, I have put you in isolation and I will order everything you eat! I hope that you will understand. My bed has been moved into the hall just outside your door, because it is necessary that I monitor your condition every two hours throughout the day and night. So don't be alarmed if I awaken you while performing my assigned duty. It might be comforting for you to know that even though you have been isolated, you are still connected to medical alert."

"Thanks Sergeant. Yes, I understand. Good night!"

As I tried to drift into a much needed peaceful sleep I let quietness become an aggravation. It's just too quiet in here for me to sleep, I thought. But compare that to what I left behind, which was hunger, pain, freezing cold temperature, loud boat noises, and roaring wind. Those things reminded me that death was always just around the corner, waiting for an irreversible mistake. Moving from that nightmare to a peaceful warm bed was certainly not about quiet surroundings, it should have been about blissful gratitude. For example: If I had run out of fuel, all of us would have surely perished. When I dropped our anchor in Japan we had about thirty gallons of diesel fuel left on board. I should be very happy and thankful to have been delivered by Almighty God from hell on Earth to the arms of Jesus. That's what I should be thinking about, I thought, not something foolish like it's too quiet! I vowed to immediately begin thinking positively, something like being blessed for where

I am with nothing to worry about except getting well! I reproached myself considerably before falling asleep while whispering over and over, "Thank you God!"

Completely unknown to me, a specialist had been flown in during the night from Tokyo. Back to the operating room I went, pronto!

The major and my new consulting physician (a Navy captain) were very busy making arrangements and discussing my history, in medical language which was, to me, no more clear than classical Chinese. But she (the captain) did say something that I well understood.

"If It becomes evident that my last-chance procedure offers no hope, I'm going to take this young man's right arm off somewhere above his elbow because we are out of time."

I was anesthetized and woke up while the captain was wiping my brow.

"How do you feel?"

I was too weak to answer clearly.

"Hush hush, sailor boy, the most important thing for now is that you woke up, we will talk later."

She began giving rapid-fire orders for my care, again in medical language which I did not understand but for that matter I was so medicated, I couldn't understand any language. The next thing I remembered was the Sergeant trying, without success, to get me to eat. I heard the captain say, "Go intravenous, he must have glucose now!"

Nothing was clear to me at that time except that people were all over me trying to save my life, a task which I had already turned over to my God hours earlier. The Sergeant told me later that I, with a smile, kept softly mumbling "Mama!" He also told me that he had never seen a medical team fight so hard and for so long to save a life. I was told later that those doctors who were assisting her advised the captain to back off because the infection had already gone too far! Her answer was emphatically NO!

The captain at that time did not remove my arm, but she began a long vigil with others helping to get my killer infection under control. As I recall, it was five days before I was permitted to sit up in my bed. Before that I could remember only bits and pieces of anything. But I felt better and decided to make jest with my Sergeant thusly, "Our superiors have put us (Ranger team) in isolation to keep us from talking to anyone, do you suppose they thought I needed a little more forgetting, because in my condition I can't remember my own name!"

The Sergeant had a big laugh and answered, "Well I don't know much about that but I do know that we have been ordered to the USS *Dixie*, which is berthed at Sasebo, Japan. As soon as you are able to travel, we leave."

"What! That's my mother ship. She is the mother hen for eight destroyers out here (Korean War zone), one of which is *Perkins*, my ship. All of our in-theater needs are supplied by her – things like medical care, dental care, fuel, repairs, food and most anything one could need. *Dixie* is a very large vessel, between 15,000 and 20,000 tons. I have many friends serving on board her, and it will be nice to see them again."

"Hold on there, captain, we will be slipped aboard *Dixie* in the middle of the night and go directly to another isolation unit. Russia is still playing her propaganda game at the UN."

"Sergeant, I'm reminded that our MWB is loaned property from *Dixie*. I wonder if she has made it home yet."

"Captain, I doubt that your boat has or ever will be returned."

My surgeon-captain walked into my room while the Sergeant and I were talking.

"Well how is my young sailor today?"

"I believe I'm gaining on it ma'am."

"Help him lie down, Sergeant."

She took more time than usual to study my records, and asked the Sergeant question after question while making her own evaluation. Therefore I became very suspicious that something good might be up concerning my physical condition. Nothing was expressed during the evaluation but "hmm" and I was exploding to know what that meant.

USS *Dixie* and her destroyers, 1951

Image: William C. Warwick, Jr.

Chapter 12

"It took the staff at this hospital, during a very heated debate, considerable time to get my approval for what I'm about to tell you," she said. "But before I start, I want you to know that hell will freeze over before I will ever in any way jeopardize the welfare of a patient. Here is what I have agreed to do. My assistants and I will, with the help of your Ranger team, escort you to the USS *Dixie*, which is docked at Sasebo, Japan. *Dixie* is not an unfamiliar vessel in my career and she has in my opinion the best track record of any medical facility over here (Japan). We will remain with you until we have completed the ten days which I originally set up for the first phase of your treatment. The chief medical officer on board *Dixie* has been thoroughly briefed about your case and will put us all in 'appropriate isolation,' which when translated appears to mean reasons other than medical. No one except my staff or your Ranger team will be permitted to move you because something as simple as a stumble could be a big setback for infection control. However, according to your current medical records, which I have read thoroughly since my arrival here this morning, and my personal evaluation, I am now convinced that (long pause) we have your infection on the run, big time!"

Moderate cheers from all of my teammates were followed by tears from the captain and me, who after wiping my tears and hers said, "Now do you understand why I have been so hard-headed about your care!"

"Yes ma'am, thank you, thank you, thank you!"

I took her hand with my almost good left hand and said, "Doctor, someday I'll hug you with both of my arms and that's a promise."

"That will be a wonderful day for both of us, young man! And now my medical team must make preparations to move you to *Dixie*!"

Joy sometimes brings tears, which are for the most part expected during sadness, but joy is the best way I know to describe my feelings that day after hearing such encouraging news about my gangrenous arm. It's not easy to live day after day wondering whether or not your arm will be amputated. The Navy Medical Corps captain who had defied everyone, including her superiors, to save my arm, will always be remembered by me with reverence every time I use that arm! Where does our country get such marvelous women!

In less than an hour after my doctor had told those in my room what to expect, things began to happen quickly. Almost before I knew it we were traveling in unmarked vehicles to a small, deserted train stop approximately five miles from the hospital, where we were put on board a very state-of-the-art Army medical train which was staffed by Army medical personnel and Army MP guards.

We traveled nearly six hundred miles south to Sasebo, a Japanese city which is located on Kyushu Island. Our arrival was planned for three a.m. so that no one would take serious notice of our coming aboard. I had slept during most of the trip because railroad train sounds had been very much a part of my childhood. They helped make me feel very much at home, if only in my imagination. My father was employed by the now forever gone Seaboard Railroad in Hamlet, North Carolina. As I was being carried aboard *Dixie*, my mother ship, once more I strongly felt as if I were coming home.

All of the mixed sounds of a U.S. Navy commissioned ship caused me to smile because I had not had that experience in quite some time. My new quarters, isolated or not, were very inviting, and even had a television. But the portholes covers remained bolted closed, so there was no way to see outside. I was very restless about not being able to do anything but I was getting a little stronger each day and I thought it might not be too much longer before I could go outside. In the meantime I intended to watch TV to learn what was going on outside of my floating hospital quarters, but the TV had been turned off. Also there were no books or magazines, so entertainment choices were zero.

The guard unlocked my door and after an ID check admitted my surgeon and my Sergeant. Again the captain began a very comprehensive evaluation of my records, assisted by the Sergeant. After what seemed to be an hour, the lengthy "yes, no, maybe, I understand, keep a close eye on this" conversation between them ended.

"Do you know what day this is?" the smiling surgeon asked me.

"No ma'am!"

The Sergeant spoke up and said, "It's the tenth day of your phase one treatment plan and your arm has healed far beyond our expectation."

"Wow! When will I be permitted to be a coxswain again?"

"We will talk about that later but first I'll be up-front and say it will not be any time soon," my doctor responded.

"Aye, aye, Captain, you have brought me this far and I'm going the full distance under your leadership, hoping that someday you will allow me to captain another boat."

"Why another one when your's is on *Dixie*'s boat deck right now undergoing full refurbishment?"

"Are you kidding me?"

"Why should I kid you about that? The Sergeant and I saw her less than an hour ago."

"When can I see her?"

"I will permit that after the interviews are completed. But for now my hands are tied."

"When will the interviews be, Sergeant?"

"I don't know Captain!"

"Well, this visit has greatly lifted my spirit and I am fortunate to have people like the two of you helping me."

"I hope you still feel that way after this cold turkey (no pain killer) examination. I will be looking for nerve and muscle responses. Sergeant, remove his pillows, lay him flat on his back and remove all of the bandages."

Chapter 13

Even though I had been warned, I wasn't prepared for the pain I would suffer during my doctor's long search to determine if there were any nerve damage indicators in my arm. Also I was not prepared for the time it would take her to find out if there were any such problems. She and the Sergeant constantly talked medical language while skillfully manipulating my right arm, always just a fraction below my screaming pain level.

This is no time to be macho, I thought. Scream and maybe they will get the message. But I didn't. Their talking back and forth about what they were doing while ignoring my agony seemed at the time to be cruel, but it wasn't, it was necessary. A little comfort emerged from time to time when I thought they had finished a particularly painful area but it was short-lived because they were constantly working all over my arm. I'm easily convinced that the same all-over strategy is used, with the same degree of intensity, when one scrubs a floor! Finally they completed the procedure, precisely at the end of my endurance.

"Are we still friends?" the doctor asked.

Due to pain and shock, I was a bit late getting out my answer – "Yes ma'am!"

"Would you like to know what we did?"

"Yes I would!"

She began with assuring me that all of the dead tissue had been removed and that the inflammation was almost gone. There were several other medical things she revealed that I knew absolutely nothing about. "Now, I'll tell you my plan for your phase two recovery, based upon what I've seen and done here today. You are, thank God, on a fast track to full recovery and within thirty days, if something unforeseen doesn't come along, I will certify you fit for limited duty."

"Does that mean assault boat duty?"

"It does not, that will take sixty days. You will remain on board *Dixie* for two months, after which you will take me for a check ride in your refurbished boat. Then I will make my decision about your returning to duty as an assault coxswain."

"I will look forward to your check ride ma'am."

They were replacing my bandages when she made that statement, and I was so excited they had to almost hold me down.

"Both of you are responsible for saving my life and my Navy career and I wish I could think of the right words to tell you both how I truly feel about that."

"Captain, I'm already indebted to you for a big hug. But to you, Sergeant, I will commit only to a big handshake."

Both of them smiled while remaining focused on a more important task, which was putting fresh bandages on my hands.

"There, how's this?" the captain asked, while holding up my right hand.

"No cage!"

"No cage, but you must be very careful for the next week."

I held her creation up for the Sergeant to see and said, "Top this if you can!"

In a few minutes he raised my left hand, on which he had made a little bow, and said, "This ought to do it."

"I'll need to think on this for a while because they both are a masterpiece."

Everyone laughed. I felt so positive about my future that I could hardly contain myself, because now the doubt had been removed and I could laugh at almost anything.

The captain looked at the time and said, "Uh-oh, I had an appointment with the admiral twenty minutes ago, now I don't even know if I have a job!"

As she rushed toward the hatch (door) she promised the Sergeant that she would get a copy of phase two for him soon. The Marine guard, the Sergeant, and I were left alone, so I raised my bandaged hands and pretended to be making medical talk.

"Anything on the interviews yet?" I asked.

He looked at my right hand and then my left as if he were complying with my concern and answered, "Nothing yet, Captain."

"Have a good day Sergeant," I called out as the guard opened the hatch for them to leave.

The guard's duty station was outside my door, unless someone remained with me in my quarters. Then he also remained. Once more all alone with nothing to do for entertainment, I began thinking about my boat being totally refurbished. That thought dominated my mind to the point that I began to imagine a number of changes I would like to add to the overhaul list. A standard U.S. Navy MWB (motor whale boat) that had so recently survived perhaps the most challenging voyage ever undertaken, on that class boat, needs no changes, and to think I would dare recommend them for that workhorse of the Navy, not hardly! Anyway, to even recommend changes to the twenty-eight foot MWB would be a blasphemous thing for a tiller coxswain to do. But it was fun to think about those things.

One I would recommend, is a trip line for the sea anchor, to prevent losing it overboard when it was too dangerous to haul it in. However, with the right size line (rope) I could do that myself. Another change that would be very helpful would be a prefabricated metal framing kit patterned to fit the boat cover when making a shelter. There is no doubt in my mind that without the ingenuity of the Rangers to build our shelter, we would have, to the last man, frozen to death. And yet another thing comes to my mind, if the Rangers had not come up with some homemade ideas to keep me from freezing to death in the cockpit, none of us would have come back because I was the only qualified coxswain on board. Also, if the engine housing had a blower vent to distribute engine heat when needed, that would protect the team from extreme cold. Fun to think about but unlikely to happen! Continuing my daydream, I began to wonder how long it would take to complete the overhaul and how long it would take for a return to Nando Island.

Chapter 14

My door opened and in walked my team leader with two civilians. "Captain these men are here to talk with you." (For the reader's information, I have been forbidden to use names of significant people, and even some places, in this writing. The only name I have used so far is Sam, the first name of my best friend and MWB engineer, whose surname I might get permission to use at a later time.)

"Welcome gentlemen, come in and have a seat. I'll be happy to talk with you."

The first of my visitors who spoke reminded me that I had requested my team leader to be present during our meeting.

"Will you share with us why you requested his presence?"

"Because I want a trusted witness present when I answer your questions!"

"What questions do you expect?"

"Well it's a sure bet that they will be about these bandages I'm wearing."

Everyone laughed.

"Well, your team leader is welcome to be with us, Captain, you might like to know that we have already interviewed several of your teammates and they made the same request. Before we start the questions though, I would like for you to know that what you and the others in your boat did was the most exciting sea adventure I've ever heard of or read about. By the way, my favorite readings are sea stories," he continued, "and to do what you did in an open-hull whale boat is nothing short of a miracle. The fact that I'm sitting here talking with you …."

"Don't forget my team leader, sir."

"Pardon my oversight. And your team leader."

"How long did it take to locate your objective?"

"Six, below-zero, nights, but be reminded sir that we could only operate at night because we were uninvited military intruders (spies)."

"How far north did you go?"

"Approximately two hundred and fifty miles."

"From where did you start?"

"I can't answer that, sir."

"How did you find what you were looking for in the dark?"

"Good logic, good people, good leadership and the blessings of God! We could have done it in four nights but an arctic blizzard slowed us considerably, and that's how I became disabled."

"Would you mind explaining to me in what way you were disabled?"

"I held up my freshly-made bandages and said, "Frozen hands and right arm."

"Who relieved you as captain?"

"No one. I was the only one on board qualified to do that job."

"Are you telling me that even though badly wounded, you remained captain for the river attack?"

"Yes, I am, because there was no other choice. And if we were destined to die anyway, who cares about frozen hands, I would have steered with my elbows if it had become necessary. Every member of the team doubted that we would survive the attack but all of us preferred to fight, as long as there was any life in us, instead of sitting around doing nothing but waiting to freeze to death.

"Allow me to tell both of you something that defies belief, let alone understanding. We were sent north to locate a very sensitive Sampan staging area that was being used by North Korea, Russia and China to transport soldiers – a few on each of thousands of Sampans, to flank our army and consequently kill large numbers of UN troops. Our team was already half frozen to death at about 20 miles south of our goal. All of us strongly told our leader that we did not wish to go home until our mission had been completed. After having gotten very close and not attack what we had faced death a dozen times to find was just not acceptable to us. After six days and nights of nearly freezing to death to reach our goal, we wanted nothing less than to show our enemy that we had discovered their secret operation base and had passed through their bragged-about impenetrable iron curtain.

"At our insistence, our leader started putting together a 'go for broke' plan of attack. He was our coach every step of the way and you can believe me when I say that without him this conversation would not be happening. It's very difficult to put into words the courage and patriotic zeal which was exhibited by eight men who believed they would soon die for their country.

"Can anyone even imagine our taking on China, Russia, and North Korea! We were determined to go out fighting for our country and I was very proud to serve with them, the likes of which are rare indeed!"

"Captain, the reason we are here is to make sure of our claim that Russia and China have been from day one very much involved with the invasion of South Korea. My assistant and I must insure that our evidence is valid and that there are no mistakes or contradictions in it. Because our claim will very soon be formally presented to the UN and must be convincing and flawless! This interview is very important, so relax and take your time to answer what I must ask.

"Did you see anything brought on board your boat either after or during the river attack?"

"Yes sir, I did."

"Tell me exactly what you saw."

"Several boxes."

"When the Rangers returned on board your boat you are saying they were carrying something other than their weapons?"

"Yes!"

"It was dark wasn't it?"

"Yes sir it was."

"Then how were you able to see anything?"

"From the fires."

"What fires?"

"The Rangers set fire to anything that would burn and after that there was plenty of light, in fact there was too much light for a covert operation."

"Did you at any time see the contents of those boxes?"

"No sir. After the boat was towed to the fish factory in Japan the Sergeant turned them over to three Rangers."

"Did you actually witness the turn-over?"

"Yes sir I did."

"How many boxes were there?"

"Three sir."

"Are you sure about that number?"

"Yes sir I am very sure."

"Thank you Captain!

Chapter 15

"There is another thing we must discuss before I leave you, Captain. You probably have wondered about the total isolation we have imposed upon your entire team since the rescue at the fish factory. The answer is that your mission to the 'frozen north,' as you call it, has unfortunately gotten your team into the middle of a big debate between Russia and the USA at the UN concerning who is responsible for starting the Korean War.

"Our country fully believes that Russia and China are completely accountable for encouraging and equipping North Korea, a country with very little military capability, to invade South Korea, but we must prove our belief before an international tribunal called the United Nations. Therefore we must sometimes use tactics that might seem to you as being intrusive and over-protective. But in this kind of business, things change quite unexpectedly and time is not always on our side when we are faced with surprising situations like the Yalu River attack.

"Our adversaries (Russia, China) well know that the river attack gave us a windfall of proof that their military involvement is the true cause of the war in Korea. And believe me when I say they will do anything to make that evidence appear to be false! And the word 'anything' in this case includes their determination to capture one or all of your team. On the other hand we must keep all of you safe in the event that we would need your help at a later time. And that means we will be involved in your lives for an unknown amount of time to come.

"Now comes the really hard part! Regretfully, I must tell you that very soon the bravest and most cunning spy team in America's military must be disbanded and its members scattered throughout our country as if they had never existed. We must make sure that none of you can be found by either Russian or Chinese agents and forced to discredit the brilliant river attack. Our country is full of Russian and Chinese spies who would be highly rewarded if they could capture any one of you and force a denial that any evidence we claim to have from the river attack is fake.

"The reason they are so fired up is because they know two things that are absolutely true: one, those boxes hold evidence that is very damaging to their already questionable international reputation, and two, those boxes did not burn in the fires set by the Rangers. On more than one occasion they have denied that the river attack even happened. But if any of their various denials are substantiated, it would totally destroy our claim that they started the Korean War and as a consequence we would have few allies willing to help us fight that war. We are working extremely hard to make sure that does not happen.

"It's easy to see that Russia/China's side of the debate is full of contradictions and false claims, but the burden of proof is heavy upon the USA. Sadly, less than a majority of the UN members are our friends and most of the others lean toward communism and will therefore be very hard to persuade to our cause.

"Work has already begun on the how and where of your team's disappearance. But a final meeting here on the *Dixie* is being planned before separation starts. There must be absolutely no effort made to contact one another for a period of time that will be clearly specified in a document that you must sign before leaving the *Dixie*. How you leave, when you leave and where you will go will be known only by the planners. Your military records will be purged to suit their needs."

Because of the Sergeant's facial expression, it was obvious to me that he already knew about what I had just heard. I was so shocked that I was unable to speak for several minutes. They left me standing alone in my room trying to take in what I had just heard, and believe me, that did not happen quickly! There were so many things I wanted to ask, but I guessed that could be done later. My whole world had suddenly been turned upside down, and they left no one with me to answer important questions about my immediate future which made me very unhappy.

For the first time in my life I was backed into a corner from which I could see no way out. Oh well, the best thing to do now was wait until all of the facts were known, and then make judgments.

In about an hour my team leader returned to my room and told me that I would not be moved until my recovery was complete. And then, to my great relief, I was told by the Sergeant that he would remain with me during that time. Later I learned that my doctor, the captain, had put up a big fuss over the possibility of my being moved because I was at a critical time in my recovery. This was all fine but to think I would likely never see my teammates again, after the final get-together, was a very sad thing for me. We had fought together on the threshold of eternity many times but to be told that my brothers at arms would be out of my life forever broke me down.

The Sergeant put his arm on my shoulder and said, "Captain, you are among the bravest men I've ever served with and yet the most tender-hearted. Remember in earlier days when I tried to teach you that casualties were something you must learn to accept? You never managed to do that but I kept you anyway because of your bravery and dependability. Do you recall those two peasants who saved your life? I knew before you began your explanation to me why you did not kill them. But I had rather have you on my assault team than anyone I have ever known because you are not only brave but you are the most skilled seaman I have ever seen. Anyway being emotional in our business is not fatal."

Chapter 16

The Sergeant and I talked for a long period of time about the ugly side of war and what it does to all of its participants, particularly the young, who range in age from sixteen to twenty-one.

Yes, I said sixteen and meant to! I'm quite sure that I have been adversely affected by unforgettable war experiences and will surely have considerable difficulty readjusting to my home life if that time ever comes. I have already missed learning many of the social skills that have customarily been acquired by most young men my age. But as a military recruit I was taught how to kill people instead, and that is about as far as one can get from a proper social education. What I have barely touched on here is a highly complex social re-adjustment issue which is wrongly passed on by the military to the discharged victim with one useless statement – 'You must on your own learn to readjust to your past civilian life.' Now that is what I consider a simple attempt to solve very troubling mental confusion. Anyway, how can anyone readjust to some-thing that was never learned, due to their absence, in the first place?"

The Sergeant was very reluctant to talk about himself. Therefore I knew next to nothing about him except that he was a clever, fearless soldier and had earned an awesome reputation as a Ranger team leader.

The door opened and there stood my doctor.

"How do you feel young man?"

"I feel as if I'm getting better each day ma'am."

"Good! Sergeant, remove his bandages and let's have a look."

They both prepared themselves for their work and then set about doing it.

Immediately I could feel their hands gently working all over my right arm, which pleased me because that told me the numbness had improved. There was pain but it was far below the scream level. I overheard the captain say, "Yes, yes, yes. Much better nerve recovery!" They moved to the other side of my bed and began looking at my left hand.

"Whoopee, look here Sergeant!"

"Yes ma'am, this one looks healed to me."

"It is! No more bandage for this one but I'll send up some salve to use on it for a few days. I want to toughen it up.

"Well sailor, it looks as if I'll get my hug and boat ride a little earlier than sixty days."

"How much earlier ma'am?"

"About a week, if that's OK with you."

"Yes ma'am, you've got a date!" I answered with a great big smile.

"But you will still need a little help until then so I will have the Sergeant move in here with you. The door guard is no longer needed because your team has gone, and since you are crew on one of *Dixie*'s destroyers, your concealment will depend upon your being treated like any other member of the crew. You may go about the ship as you please if the Sergeant is with you."

"Ma'am did you say that my team has gone?"

"Yes I did, but I do not know anything beyond that."

I was terribly disappointed but I had nothing further to say except "Thank you, ma'am. Sergeant, let's take a turn around the *Dixie*'s main deck (1100 feet)."

"Give me a few minutes to move my things and we will do that," he responded.

It was such a beautiful day and the outside air was so refreshing! The Sergeant spoke up and said, "I do not believe it is necessary for me to say what I'm about to, but I need to get this totally off my mind so here goes. We must never again for any reason talk about our teammates or our final mission with them."

"Consider it done, Sergeant!"

Suddenly we were stopped by a hospital corpsman, who pinned an ID tag on my clothing and said, "I was on my way to your quarters to do this. You are Captain [deleted] patient aren't you?"

"Yes I am, thank you."

We continued our walking for a while but it was becoming a bit much for me. I had not exercised my legs in a long time and my stamina was less than I had anticipated.

"I guess we had better go back, Sergeant, I'm feeling a little weak."

"Sure thing Captain! I'll get a wheelchair if you like."

"That probably is a good thing to do."

"Sit here on this hawser cleat until I get back."

While I was sitting there, I decided that it was time for me to begin an exercise program. When the Sergeant returned I asked him if the *Dixie* had a rehab facility.

"Yes they do, and if you feel ready I'll make the arrangements."

"Thanks, I have come to realize that I must eat well and fight hard if I expect to recover my strength in time for the check ride. And only you and I know how badly I want to get my boat back! Also I would like to start having my meals with the crew."

"That I will arrange also, Captain."

"Thanks again, Sergeant! Come to think of it I am a crew member of *Dixie* and do not even know my way around on this monstrous floating city. My commodore's flag is aboard this vessel and he commands eight destroyers that are assigned to her, one of which – *Perkins* – is mine."

After we had returned to our quarters the Sergeant said, "I'll be back in a few minutes," and left. When he returned he said, "We are going to have our evening meal in the mess hall."

"It might be interesting at this time to mention that everything on board *Dixie* is on a twenty-four-hour schedule in three shifts. She really is a floating city, which keeps her eight destroyers ready to fight and their crews ready to do the fighting. I just can't pass this one up, she has a great MWB repair facility with a well-known reputation for quickly repairing damaged assault craft. Her inventory is twenty boats – sixteen on active duty with the destroyers, and four standbys.

"Now, I'm going to get busy and repair myself so that when my boat is ready to be launched, I'll be ready to resume being her coxswain."

Chapter 17

My doctor is a friend but she absolutely will not pass me for unlimited duty as an assault coxswain unless I'm physically qualified! No favors here, she is just not that kind of professional woman. Having wised up to that fact, I resolved to try being more independent and as a start I wanted to feed myself the evening meal. I discussed that intention with the Sergeant.

"That's good to hear but you must remember there are some things that you must not do yet, such as lift anything with your right hand, or get it wet. But this I promise, do as I say and I will have you physically ready to pass the check ride!"

Thanks Sergeant, that relieves me very much."

"Well, I haven't told you this but I have another assignment and I have requested to keep you for my boat captain, if I can get you well soon enough. And furthermore the *Perkins'* captain has already consented for you to remain with the Rangers."

After that news I was stunned, happy, honored, and who knows what else. "Sergeant I'm so pleased with what you have just said that I do not know what to say."

"Well don't say anything, let's go to the mess deck and eat. Now remember, don't get too independent because an injury could set us back!"

"You are my trainer, and I will do exactly as you tell me."

The mess deck was alive with sailors and Marines talking and laughing while they ate, and that made me feel right at home. The food tray was too awkward for me to handle with one weak hand so the Sergeant cautioned me not to try. A mess cook overheard him and came forward to help me. After we had gone through the chow (food) line he asked me where I would like to sit, and proceeded with my loaded food tray to the place I had designated.

"Thanks mate!"

"No problem, hope you enjoy your meal."

Without any help, I ate everything on my tray! For the first time since being hospitalized, I was firmly focused on reality. With unwavering determination and the Sergeant's help I will once more captain my boat. Some changes to it, that had been dancing around in my head for some time, had started to dance again. It wouldn't do any harm to recommend, I thought. Anyway I have a plan in mind that would without a doubt improve that already famous design. Maybe in a few days I will visit the boat maintenance shop and test my acceptance before I mention anything about changes.

When the Sergeant and I returned to our quarters the television was showing an old movie from *Dixie*'s film library through its broadcasting facility. The same was true with the radio. Both services were available from reveille (6 a.m.) until taps (10 p.m.). Since I had no interest in either, I gathered up things I would need and headed toward the assisted bathing area, which was part of the hospital's out-clinic. They had a device which would fit over my arm to keep it dry. After that contraption was deployed, I could bathe myself quite well.

I knew the Sergeant had a big day planned for tomorrow, therefore I turned in early to prepare for life after reveille. Just as I had expected, he started his routine before the reveille bugle call had silenced. Very quickly I realized that on this day I must do everything for myself, even my buttons, with only one exception – the food tray.

He had already laid out my uniform for the day, which was not hospital white but chambray work clothes like everyone else wore. Thirty minutes after I had put my feet on the deck, I was dressed and on the mess deck having breakfast. After that it was back to our quarters to review his plan for the day, which included our muster (roll call) with the boat repair people on the boat deck for morning colors (American flag raising) at 8 a.m. Needless to say the Sergeant was preparing a way for me to become friendly with them, and perhaps satisfy my unending desire to see my boat. And who knows, maybe they would at least listen to my improvement ideas.

Prior to leaving for the boat deck, the Sergeant fixed a neat sling for my arm – so neat that it was tied to my belt, therefore I could not move it. After roll call the Sergeant explained our presence to the leading petty officer. He responded that we could muster with his group each morning if we wished to do so. There are near two thousand people at any given time on board *Dixie*, and that is why morning roll call is so important!

The Sergeant had successfully put me where he and I wanted to be – on the boat deck. The next step would be up to me. After colors there were quite a few welcome-aboard handshakes and introductions, but no invitation to visit the boat repair shop. But there was plenty of time and I didn't want to push it. We left the boat deck and went directly to the main deck and began walking. After that it was off to the rehab gym for tolerable calisthenics. We stopped for a short break and to my delight I was holding up unexpectedly well. I guess my positive mental attitude had a greater influence on my stamina than I had anticipated. And the last exercise before our noon meal was slow walking on the boat deck.

Now I'm sure I don't need to explain the reason why the Sergeant made that choice. There were several hand waves directed at us but nothing beyond that. But just as we were about to leave, one of the boat shop workers waved at me and started walking in my direction.

"Are you crew from eight-seven-seven (*Perkins*)?"

"Yes I am," I answered.

"I kept looking at you through our office window and told my mates I believed that you are the coxswain who motored our boat across the Sea of Japan!"

I did not respond to that!

"We are working on her now, would you like to see what we are doing?"

Does a cow give milk? I thought. "Yes, I would!"

The three of us walked together into the very large shop. "There she is!" our guide said, pointing.

Chapter 18

"May I move in closer?"

"Get as close in as you wish, Captain!"

Others had begun to gather around the coxswain (me) who had done what they had believed impossible with a MWB. They all seemed to know, from the boat's serial number and from my signature which could be easily confirmed by the ID which had been pinned to my shirt pocket, that I was the coxswain who had done the impossible. Also there were records that would clearly proved I had signed out, from their boat inventory to *Perkins*, the boat at which we were all looking. And that's why the sailor who approached us asked if I were crew from eight-seven-seven. The answer to that question positively identified me as the coxswain who had crossed (400 miles) the very cold northern Sea of Japan with their boat.

They also knew that their boat had been recovered at the fish factory in frozen northern Japan where we had left her. The Sergeant realized that I was getting into trouble by not answering some of their questions so he told me to tell it all except the river episode. "What river episode?" I asked.

"Never mind Captain!" He winked at me and continued, "Maybe I should explain to everyone here that there are things about our previous mission that we cannot talk about and I hope that all of you will understand our refusal to answer some of your questions."

Everyone agreed and said they understood. But one, who appeared to be the leading petty officer (chief) in the boat repair shop, informed us that he had been told to repair my boat under cover so that it could not be spotted from a fly-over.

"The Russians do not want the boat – they want the Sergeant and me," I said. "And if they locate our boat, they will sooner or later find us and then we are as good as dead. So please do not help them find us by telling the rest of the crew we are on board *Dixie*. Talk among this large crew (1700 plus) will travel far in a short period of time."

The chief spoke up and said, "I figured it was something like that. But don't either of you worry one minute because when I've finished some design changes, no one will ever know it is the same boat that was left at the fish factory. But you can believe it was photographed from one end to the other after you left. And you can also believe that every last one of those pictures is now in Russian hands! Therefore, they well know how it looks. Knowing that, I'll make design changes, such as a new serial number ID plate, and then throw the old one over the side. All of my records will be changed to protect you both, but we will talk about those things later."

I almost dropped dead after I heard the words "design changes." Immediately after hearing them, I asked the chief to have noon chow with us so that we could continue our talk.

"Thank you coxswain, I accept your invitation."

Away we went to the mess deck. The chief immediately became more than a little interested in my injured arm, and that was when I knew we had his sympathetic attention, which was good for us!

"Sergeant, I will appreciate your telling him (Chief) about my injury."

"Yes I'll do that Captain!"

He started, and kept it up for at least an hour. The chief was interested in every detail about the boat's performance, but my being frozen to the tiller was something very special that completely mesmerized him. The Sergeant did a very thorough job telling him about our many life-threatening experiences and how we had done such unusual things to survive – right up to the fish factory. However, I was very impressed with his skillful avoidance of the river attack. Even though the "AP" news had been filled with what he didn't believe a MWB could do, we had done countless other things to stay alive that were equally unbelievable, so I expected at any moment the chief would ask if we had made the river attack. But perhaps he remembered that we could not answer that question, even though I'm sure he believed that we had done it.

I was ready to jump in but the chief spoke up and invited us to return to the boat shop with him.

"Yes, we will do that," I answered.

And thus began the building of the most versatile and certainly the most famous assault boat in the U.S. Navy. When I was all done using her, she became a permanent fixture on *Dixie*'s boat deck because the Navy would not abandon its historic MWB, and they never again, to my knowledge, put the new design on the water.

But what I didn't know at that time was that the Rangers had plans to send me back to the frozen north. Therefore, the changes were allowed for one boat only. Back at the boat shop we started with the boat cover being used as a shelter to protect the crew and passengers. The chief sent for his sailmaker. The Sergeant explained to him how he had used the cover that we had, to build a temporary shelter to prevent our freezing to death and to help keep water out of the boat during severe weather conditions, which was all the time. That adaptive capability accompanied by larger bilge pumps would greatly reduce chances of losing the boat and its occupants before even reaching their target. Also prefabricated removable supports or the like should be made so that the cover could be quickly put up as a shelter, or taken down and stowed away.

"Sergeant please tell the chief how you got heat from the engine."

He began with telling how he removed the top from the engine housing so the heat could rise up and away from the hot engine area. The chief stopped us and requested that we write a list of the changes we would recommend for future operations.

"I promise that I will do them all but I need to prioritize them so that my people will be better organized for their work."

Chapter 19

"Thank you, Chief, but may I ask one question before we leave?"

"Certainly, what is it?"

"The engine in my boat is a Gray Marine 120, as you well know. But with our mission's heavy supply needs it is necessary for me to run it at 75% throttle or more all the time, which creates too much strain on the engine and also too much noise for covert activities. I'm sure you understand that in our business, noise is not a good thing! With such a heavy load, I can barely head-up even in moderate seas, and her bow will not properly plane even at full throttle. Is it possible to put a more powerful engine in my boat?"

"Yes it is. I have a Gray Marine 200 in my inventory that will fit nicely in your boat," he responded while writing. "It will not be necessary to put the engine on your list, I have it on mine. By the way, if you had stopped your engine you would have been unable to restart it because there was probably less than an hour left for it to run. The pistons are cracked and the rings are badly worn from lack of oil. I can't imagine how your boat ran full throttle for thirty-six hours without imploding! But permit me to say that God has spared both of your lives, and those with you, and only He knows why! However, it is my intention to make the odds much more favorable for you this time. Also it's my intention that your boat will be my best overhaul yet, but to do that I will need continuing input from you and the Sergeant."

"Thank you chief, because this boat will be our lifeline and we will spend as much time in your shop as you require."

I will always believe that he knew we were going back to the frozen north and wanted to do everything possible to improve our chances of coming back. "Better this time" was to me a clear tip-off! How did I know that? Because everything was too easy and that's not the way my Navy usually does things! It was as if word from some high authority had gotten to him (Chief) and ordered that he give us everything we requested.

"Thank you, Chief, we will see you in the morning at muster!"

"Don't forget to finish your exercise!"

"We won't! Have a good evening," I said.

The Sergeant laughingly said, "That was easy! If we had asked, the chief would have put a shower on our boat."

"Yeah, I agree, but I'm very suspicious about how easy it was. It's my guess that we are going north again and this time for more than a week!"

"Well, things will start churning up in a few days and then we will know where we are going. In the meantime let's just wait and see how the chief reacts to our 'want list' tomorrow after roll call."

"Good advice, Sergeant!"

We decided to complete our rehab schedule before eating. While we were slow walking the Sergeant told me that four replacement Rangers would be sent to him in a few days for either his acceptance or rejection.

"That made me believe that we have a very important mission coming up that will be entrusted entirely to my choice of team members. Do you have any people on *Perkins* you would like to have for your engineer and lookout?"

"I had the best, but I need to think on that a while. We both know how much time we have spent training our people and how long it will take to train replacements."

"You sure are right about that, Captain!"

During our evening meal I wrote two names on a paper napkin, handed it to him and said, "Either one of these will do for our replacement engineer. Both have dependable service records and can be trusted to keep their mouth shut! I'm still thinking about how to find a wild, courageous, well-trained lookout but so far I've drawn a blank."

"Your choice is my choice, Captain. I'll take care of this one (engineer) right away."

"Come to think of it we need our engineer as soon as possible so that he can be involved while the larger engine is being installed. But I want one thing to be clearly understood by all new Navy members of our team. You, Sergeant, are absolutely the commander of this team and I want it said in my presence so they will not try to lean on me for protection. Navy people are sometimes a little hard-headed about obeying ground force commanders."

"Are you suggesting I do to them what I did to you!"

We both laughed as I said, "That I am, Sergeant, that I am! It's a must that we know their grit before going into action with them."

We left the mess deck and decided to walk around the main deck again. During that walk I recalled and mentioned to him a fearless, outstanding lookout whom I had trained a while back but I thought he had left the Navy.

"Would you like to have him?"

"Indeed I would!"

"I'll try to find him. When we are back in our quarters write on this napkin everything you can remember about him, and if he is anywhere in the military service I'll find him."

Back in our home away from home we began composing the want list. We had mentioned the larger engine to the chief before we left the boat deck, and that alerted my mind to fuel storage.

"We need some kind of racks to prevent the tins from being thrown about on the deck. That would also give us more living space. We also badly need some way to dispose of the empty fuel tins for the same reason, deck space. I guess what I'm getting around to is, on a long deployment some consideration should be given to creature comfort and a helter-skelter cabin badly interferes with our mission." The Sergeant was writing while I was talking. "Everything possible should be done to protect us from the bitter cold. That should include a survival suit for every man. The brass tiller must remain because it will not break like one made of wood, but it must be insulated to prevent what happened to me again!

"Food for a week or more should have more variety than K-rations."

Chapter 20

We completed our twenty-two item list just before midnight. We also believed that we had covered all areas of need and that the chief would consider changing or repairing at least half of them, assuming they were worthy of his attention. As always, the Sergeant was up before reveille and began calling for me to hit the deck as the very first sound came out of the bugle.

"We have a long day ahead of us so roll out!" he said.

Rolling out of my bed was easy but awakening took a little longer. Before I knew it we were on the mess deck having breakfast. By the way, the same mess cook continued to help me with my tray every meal and for that I was very grateful. During our delicious meal, I began to be aware of tingling feelings in my injured arm – not discomfort, just feeling alive as before being frozen. I alerted the Sergeant about what was going on with my arm and he said, "It has begun to come back and that is good. Are you sure there is no pain?"

"Yes I'm sure."

"To be on the safe side, we will check with captain [deleted] after roll call.

The chief already had plans for our workday, however he understood my medical necessities. "I expect that we will return soon," the Sergeant said.

"Must you go also?" the chief asked.

"My orders from doctor [deleted] are to remain with him until she has released him."

"No problem, I'll hold up on your assignments until your return."

"Thanks Chief, we won't be long!"

"Good morning! I'm glad to see both of you, do you have a problem?" The Sergeant did the talking and explained to her what I had told him.

"Well let's have a look. Young man you have reached your final stage of healing and everything looks good to me. No pain?"

"No ma'am, no pain!"

"Sergeant lighten up his bandage, I would like for a little more air to get to his skin."

"How is this ma'am?"

"That's real good. I would like to see him every other day during this final stage. And also I will need to see his rehab reports at that same time."

"Yes ma'am."

"Sergeant I have good news for you," I said as we stepped out onto the boat deck.

"Well tell me about it, Captain!"

"I will be your team's boat captain on our next mission to wherever!"

"Well that's great to hear, I thought I might end up with someone who could get me killed."

The chief met us and said, "That didn't take long! Sergeant, I would like for you to work with my sailmaker, and Captain, you will be working with my deck hardware man. Oh, I forgot to ask, did you make the list?"

"Sorry, I almost forgot," the Sergeant said, and gave it to him. As he studied what had been given to him we stood fully prepared to debate any item on it. Finally he spoke and asked, "Have either of you been given a departure time for your next mission?"

"Not yet," the Sergeant answered.

"Well, as I promised, I will do everything on this list and more, but I must get some kind of time frame for my boss so that I can schedule production within the time I'm allowed."

"Thank you Chief," we both said.

"All right, let's put you men to work."

He took each of us to our respective work stations, introduced us, and left, saying, "Have a nice day gentlemen!"

My work mate asked, without delay, "What do we tackle first?"

"My compass!"

"What's wrong with it?"

"I can't use it!"

"Tell me why!"

I explained that what I did had to be done during darkness and that if I have any light showing, I become a target. Therefore, all DC light service wires are cut to prevent being accidentally turned on. Navigating under those conditions is extremely dangerous and presents a high possibility for getting killed.

"Well, we must solve that problem! Let's kick around a few ideas. I need for you to be in your boat."

He hand-signaled for someone to help him. When I had once more stood on board my boat where I thought I would never again stand, my feelings of joy and love for an inanimate object were so powerful that momentarily I could not speak. Those who had helped me climb aboard could easily tell what was happening to me – they had seen it many times before.

My co-worker gave me plenty of time to look and feel around.

"She saved my life you know!"

"Yes Captain, I know."

Finally we began trying to solve the compass problem.

"Where do you usually sit when you are piloting your boat?"

I pointed.

"Sit there so that I can take some pictures. The compass is satisfactorily located for daylight motoring, right?"

"Yes, that is correct."

"So what we need to do is create some way you can see it at night without being blown out of the water."

"Exactly!"

"Now that we have identified the problem, let's start looking for a solution. When you are motoring in enemy waters, I assume that you are sitting as you are now instead of standing."

"For the most part yes but if I'm standing, I have the target in sight and then I don't need the compass."

"Okay, let's build a flexible periscope with a night vision light that cannot be seen from outside."

"That sounds like a winner to me!"

"After I've done some measuring, we will go to my work bench and build a working model out of cheap material which is easy to modify, and then test what we have. But you stay where you are until I put together some way for you to get on and off your boat."

It took time but he and a co-worker made a very safe way for me to embark and disembark. His very large workbench was every craftsman's dream. I could hardly stop looking and asking question after question. Everything appeared to be in its place and efficiently arranged for quick access. It was at that moment I knew that if anyone could build an apparatus to solve my compass problem, he was the one to do it.

Chapter 21

He began drawing on paper, from his measurements, a profile plan that would soon become his construction model. Watching him work was like watching a master of any discipline use his unique way of achieving perfection. He concentrated on what he was doing with very little talk. However, he did ask me now and then how the plan looked, to keep me from feeling as if my presence were an unnecessary intrusion. When he had finished the drawing, he asked me to have a look before he began cutting material for a temporary periscope model.

"I'll consider placement of the mirrors later," he said, "because they will require an adjustment capability."

"It looks great to me but I do have a question."

"Fire when ready Captain!"

"Will the peephole be watertight?"

"Yes but I'm still thinking on that problem. Maybe you can help on that one."

Bingo! He had found something for me to do so I would not feel so useless.

"May I bother you for a scrap of plywood and a carpenter's pencil?"

"Coming right up Captain, you will also need a square and compass."

"Yes I will! If one puts the base of the periscope stand directly over the compass box, a forty-five degree angle for the peephole should do nicely."

I had just begun drawing that idea on the plywood when mess call was announced over the PA (public address system).

"You may use my sink to wash your hands if you like. Sorry, I'll help you, follow me."

After I had without any help used the facility, he washed my hand, covered his with a towel, and said, "Shake!"

The way that he had helped me considerably reduced my feeling of dependency. Before we left for the mess deck my Sergeant walked in and asked, "Are you ready?"

We both knew what that meant. "Yes I'm ready."

"Okay let's go!"

My work mate customarily ate his meals with the boat shop crew so they could discuss things that only concerned them. The Sergeant and I proceeded to the mess deck, excitedly talking about what we had seen happening to our boat.

"If we are sent north again we will be much better prepared to endure it. You will not believe what the sailmaker is doing about making a shelter with a new and larger boat cover!

"And you will not believe that we will have a lighted compass that we can use at night!" I added. "I sure hope that we will have enough time to complete everything."

"So do I," the Sergeant agreed, "but I really don't believe there will be. That is why we must choose to do the most important things first. And we must not forget that we have six new replacements to train."

As usual the mess cook was waiting to carry my tray to my chosen place, but on that day I would have had trouble carrying it with both of my hands – it was really piled up. However, I was very hungry and I ate everything on it except the chicken bones. After we had finished our meal, we decided to walk about on the main deck. During our walk we heard on the PA, "Sergeant [deleted] please report to the quarterdeck watch officer." As always the message was repeated.

"What could that be!"

"Keep walking, I'll catch up with you later."

The quarterdeck is part of *Dixie*'s main deck and is the place where anyone who wishes to come aboard her must do so on an outboard ramp (gangway) at that location. There is always a security watch to make sure that no one comes aboard without a proper ID and orders to do so. On most U.S. Navy vessels the quarterdeck is located midship on the main deck. The boarding ramp can be located, depending upon how the ship is moored, on either the port or starboard side. My walk carried me around the stern of *Dixie* and as I continued up the starboard side five Rangers approached me, one of whom was my Sergeant, who spoke first.

"I figured to intercept you by going the opposite direction. Captain, these Rangers are candidates for our assault team."

I welcomed each of them with a handshake and spoken words of welcome.

"Tell the chief I'm busy finding quarters for them and will further explain things to him in the morning at roll call."

"Yes, I'll do that Sergeant!"

I finished my walk and returned to the boat shop, where I approached the chief and reported the reason for the Sergeant's absence.

"When do you expect your engineer?"

"I wish I knew. Not many people will volunteer for the kind of work we do and we certainly don't want a draftee who would likely get us all killed!"

"Well the reason I asked is that I would like for him to be here when I repower your boat."

"Oh, I forgot to tell you that the Sergeant will be at morning roll call. Maybe he will have some good news by then."

"I hope so because I have been given a ten day max time frame and I wanted to share that with both of you. But if we all work long hours we can finish the job on time. Every one in this shop is dedicated to completing your boat before she leaves our shop. We have an idea where you are going and we are determined to give your team the best possible chance to survive and come home again."

"Please thank everyone for us! That commitment could very well save our lives! I'll be at my work station if you need me."

"Okay, Captain, I might drop by in a little while to see if you need anything."

"Thanks Chief!"

My work mate had almost finished cutting the pieces for the first working model. After greeting each other, we returned to our pre-lunch projects. He had his put together in about two hours and asked if I were ready to put it in place for testing.

"Yes, mine can be done later!"

He requested that I sit in my normal piloting position, placed the model in place and began almost endless adjustments. He was a real perfectionist! However, for that instrument (compass), that was exactly what I what needed.

Chapter 22

"How is this?" he asked.

"Very good!" I responded.

Then he wanted to have a look at my peephole drawing. "This will do just fine. Let's start cutting the permanent wood patterns."

"Okay, tell me what I can do to help."

"The first thing I want you to do is carefully listen to what I'm about to say! You have been and will continue to be good company and an excellent working partner but I am responsible for keeping you safe from all harm while you are in my shop. Because if by whatever circumstances you become hurt again, especially your already disabled arm, my co-workers would probably throw me over the side. So beginning right now you are restricted from doing anything that might cause further injury . Please don't misunderstand me, you can be lots of help to me if you will always wait until I to tell you what to do. Professionally speaking, I will tell you when I need someone to hold the other end of the board. You will notice that I have put a two-gallon red bucket on the table beside the sink. It's filled with soapy water to wash your hand. It's easy to do – just put your hand in the water and slosh it around. Your towel is hanging on a wooden peg just above the bucket. Understood?"

"Understood!"

"Now let's go to work. Come with me. This is teak wood, which we will use to build the compass periscope. Take three pieces, one at a time, from this storage bin and put them on my work bench, while I set up the saw."

One piece! I can lift more than that, I thought. But I'll do as I'm told.

And so it went until all of the pieces were cut and assembled.

"Okay let's try it again!" he said.

It fit perfectly. He was ready for a DC electrician to install a night vision compass light. At his work bench, he began finalizing for the periscope adjustable mirror locations, forty-five degree flexibility, and my waterproof peephole.

The sailor-craftsmen in this boat shop are nothing less than amazing, I thought. A shop filled with perfectionists, now that is something rare to behold!

Mess call alerted me to the fact that the morning had gone by rather quickly. The Sergeant and his Ranger replacement candidates came to the boat shop and he asked, as expected, "Are you okay?"

"I will be in a couple of minutes."

I climbed off the boat with my work mate close enough to break my fall if that happened and quickly completed a head (toilet) visit all by myself. Then I swooshed my hand in the bucket and dried it with the provided towel which I shook at the Sergeant, after which he gave me an "OK" salute. I replaced the towel on the peg and started walking toward them.

I had begun on that day to feel better than I had since before the blizzard. I had learned the vital importance of independence that my work mate, by his example, had unknowingly taught me.

When I had gotten close enough to them I said, “Well, Sergeant, now your team has a boat captain who has a compass that can be used at night.”

“I am very pleased to hear that but I must confess, I had strong doubts about that one,” the Sergeant said. “Follow me gentlemen, it is time to eat!”

While we were having our noon meal we shared many things that helped us become better acquainted. I did not ask, but I didn’t believe the Sergeant had made his decision about the replacements becoming members of our team. But there was one thing I did know, he would not be rushed. Neither would I about my Navy replacements. However, not having an engineer had become critical since we now had a deadline, about which I had not informed the Sergeant. I was concerned that the Navy might send a person who was not qualified, just to get the position filled. An engineer on a liberty boat is one thing but on an assault boat it’s quite another, especially if the candidate is told what will be expected of him.

After our meal we all went to the main deck for a bit of chit-chat. And then guess what, a PA message silenced us. “Will Sergeant [deleted] please report to the quarterdeck watch officer!”

“I hope that is what I have been praying for.”

It was! I saw him approaching us with Sam, our one and only engineer. I was very grateful to greet my best friend. We delayed too much talk until Sam had been introduced to the others.

“Since Sam has been with us from the beginning, let’s put him in with us, Captain.”

“Yes let’s do that Sergeant. I need your permission to do something.”

“What is it?”

I would like for Sam to put on his work clothes and then you and I should take him to the boat deck and introduce him to the chief.”

“Good idea, I know that the Chief will be very pleased to meet our engineer at last.”

"Yes, they have been waiting for him before setting the new engine in place."

"Permission granted Captain! I'll wait here for your return."

"Thanks Sergeant, Follow me Sam! Where did they send you after you left here?"

"Guam! I was an auto mechanic tucked away from everybody. I was very pleased when I learned that you needed me and wanted me back."

"Well Sam, needed is an understatement. It was you or no mission because no one would volunteer and even if they had the Sergeant would not have accepted them. I think the word got around and no one wants to serve with us because they believe we will surely be killed. In hindsight, I believe the decision-makers realize that breaking up the best assault team in the Korean theatre was a bad mistake because they now know what is ahead for us."

"What happened to Hawkeye?"

"Didn't the Sergeant tell you?"

"No he didn't."

"He found him in Key West, Florida. I believe they are going to bring him back also."

"That will be great! We have been together so long and I was truly uneasy about training another Navy crew."

Chapter 23

"Okay I'm ready, let's go meet the chief. Sam, your new engine is a Gray Marine 200."

"What! Will it fit?"

"The chief told me that it will fit nicely."

"Well that sure will be an improvement over what we had. I struggled with that engine the last six months of its life."

"There are several new things planned that you will like. Be advised though, they are giving us anything we want, so ask and you shall receive."

"Well that's not the Navy I have been accustomed to. Before, I would almost have to fight for a screwdriver."

We rejoined our leader and the candidate Rangers where we had left them and together headed for the boat shop. The chief saw us coming and walked over to meet us. Believe me he was very relieved to see us with, he hoped, our engineer. Absence of an engineer had been holding up engine installation as well as several other jobs on our boat. He told us all something that he had already told me, which I had forgotten to inform the Sergeant, that the boat overhaul had to be completed within ten days.

After only a few niceties the chief led us toward our boat where the uncrated engine was waiting. In no time at all, several boat shop workers came over to meet Sam. And guess what! The four of them immediately had a professional pow-wow and began work. The chief was so pleased he could hardly stand still, he was all over the place checking this and that. My co-worker had also been held up by the engine's delayed placement, so he happily joined the group with Sam. All of them jumped into the boat and began making ready. After about an hour one of them cranked the lifting machine. The chief, Sergeant, Ranger candidates, me, and a number of others cheered when the lifting machine began raising the engine.

There were a large number of things that had to be dealt with before the engine could be lowered the last few feet to its permanent location. But before the end of that workday, the long-awaited new engine's bilge bolts were tightened in their intended place. Two very interesting things had happened that motivated the engine being so quickly set. One, the pent-up frustration of workers who had been blocked from doing other things on the boat that needed to be done; and two, the leadership of Sam, the best small craft engineer in the Navy.

It was almost quitting time (1600) but the craftsmen were busy walking all over and around the boat, pointing and planning tomorrow's work. My co-worker was right in the middle of it all so I left the shop with the Sergeant and the candidates to begin my rehab without telling him. Sam knew *Dixie* quite well and would have no problem finding his way.

The Sergeant put me through a rough routine that day. However, I was pleased to learn that I could do so much! Finally, he released my right arm from my belt and coached me through thirty additional minutes of calisthenics that were specially designed by him for my very weak right arm. During our cool-down walk, curiosity pulled us back to the boat shop. The workers were still at it. I don't believe they realized that we were watching. The entire boat shop crew seemed to be fired-up about my boat's repair and its being back on the water as soon as possible.

They must know something that I don't, I thought.

The Sergeant decided against asking Sam to eat with us because to remove him would cause all of the others to quit for the day. So we all with our restless appetites proceeded to the mess deck. The mess cook as usual helped me, and put us in the same place. The candidates seemed to have relaxed a little, but not enough for me to believe they had passed the Sergeant's merciless final test. He really was a do-or-die Ranger. When he gave an order he meant it!

Back at our quarters he said the same thing to the new candidates that he had said to every member of our team. Most of what he said in the beginning was a step up to The Big One.

Here's how he started on The Big One:

"All of you (candidates) must learn that I will tolerate nothing less than absolute and immediate obedience. There will be no hesitation and no questions. If either happens I will shoot whoever did it dead in their tracks without a moment's delay because that would be disobedience in time of war. And you had better believe I have the authority to do it because where we are going the slightest delay for whatever reason can get the entire team killed. And always remember that nothing and no one is more important than our mission. Now I want three of you to go outside and close the door."

The Sergeant sat across the table from the remaining candidate. He laid a Colt model 1911 .45 caliber service pistol on the center of the table. The firing pin had been removed but the candidate did not know that. "Ranger," he asked, "are you willing to join my assault team, knowing that if you fail to obey my orders instantly I will kill you?"

"Yes I am Sergeant!"

"Pick up the weapon you see before you and carefully examine it."

"The weapon is ready to fire Sergeant!"

"So you are telling me that the weapon you are holding is ready for killing."

"Yes, it is ready."

"Point it at my heart. Now carefully cock it. I'm ordering you to pull the trigger NOW!"

He did without a split second delay.

Of course nothing happened because there was no firing pin. The Sergeant stood up, took the Ranger's hand, and welcomed him to our team. The same pledge of obedience test was given to the other three and they were also accepted as team members. One more to go, I thought, but he will not require a loyalty test: Hawkeye. We sat around talking and decided among ourselves that it would be best for the team to move in with the Sergeant, Sam, and me, for security reasons and planning purposes.

Chapter 24

There are ten hospital beds in this compartment which our former teammates were, at least in part, using because they would not allow assault team members to move in with non-team members for security reasons.

Sam came in and was very excited about his powerful new engine. The Sergeant brought him up to date on our new team members and our decision for all of us to bunk in one secure location. Sam went to each new member, shook his hand, and welcomed him. But he had a real hard time getting his new motor off his mind. "Sergeant, I can give you hull speed (14mph) forever with that engine. The only limit will be fuel. Keep that in mind Captain."

"Indeed I will Sam."

The Sergeant gave Sam a copy of our twenty-two item want list along with his permission to change or add anything he wished. According to Sam the large engines are quieter and are customarily used only for an Admiral's barge, and our having one had made him suspicious that we must be going some place very special to have such an honor.

"Yeah like Vladivostok, Russia," I added.

"How do you know that, Captain?" the Sergeant asked.

"Just a guess!"

"Well you have hit the nail right on the head because that is exactly where we are going. All team members are hereby from this moment on reminded of their secrecy oath. We are brothers sworn to be one for all and all for one. When you let anything about our activities come out of your mouth, you may have condemned your brothers to death. So think before you open your trap. My recommendation is do not speak to outsiders unless it is necessary and even then be very cautious. Stay in small groups because people are less likely to approach a group for conversation. Yes, we made the Yalu River attack and yes, the Russians are still looking for us so be very careful with whom you talk. The enemy (Russia) has recently been putting floating mines into the Sea of Japan somewhere near Vladivostok and they are drifting south and causing big problems for our Navy. Several ships have already been lost because they collided with one of those floating nightmares. Our job will be to discover where and how they are launching them. Since we travel only at night, we must be extremely vigilant or one of those mines might find its way to our boat. However, if one is sighted we must make sure that it is destroyed."

"How do we spot a floating mine at night?" one of the new team members asked.

"Good question! We have one more position to fill on our team – a lookout who served with us on our last mission. We have nicknamed him 'Hawkeye.' He will return to *Dixie* tomorrow."

He looked at me and smiled. I looked at him returned his smile and said, "Thanks Sergeant."

He continued, "He is truly the only man I've ever met who can see like a cat – at night. He will teach all of you how to be excellent lookouts. And if we will be motoring in the same waters with dynamite that can easily sink a ship, I believe all of you will be very quick learners. Some of the things you have heard me say this evening will likely be said again. But there is one thing I said that I hope is buried so deeply in your mind that you will never forget: you are responsible for my life and I am responsible for yours, we are brothers! That's all for now."

Our leader approached me and said, "Captain I need to have a look at your arm. I pushed you a little hard today during rehab exercises." After removing the bandage he said, "Captain this looks very good to me. I will put on a fresh bandage but I believe the doctor will discontinue them tomorrow."

"I hope so because I'm sure tired of being dependent. Tell me about our lookout."

"He has been hiding at a small Navy installation in Key West, Florida. They checked me out well before they allowed him to call me, and he is anxious to return to our team. They will fly him to Sasebo tomorrow and escort their passenger, in handcuffs as a disguise, to *Dixie.* After the quarterdeck clearance, he is mine. I don't have a clue what time he will arrive, I'll just stay alert for the PA call.

"You know Sergeant, I don't believe there's another man on Earth who can do for us what Hawkeye can. We sure have been blessed today!

"Yes we have Captain! I hope it isn't too late for me to shower at the rehab facility. I don't believe it is, because I'm almost sure they work three shifts."

Immediately after I got into my bunk I drifted off to sleep, before I had finished my evening prayer. It occurred to me while I was rushing to get dressed after being called, that I had not yet heard the reveille bugle. Our leader had gotten us up early, as always, to have our team's morning prayer and Bible reading. He believed that where we were going we had only a slim possibility of coming back. But he wanted all of his men to be strong Christians and stand tall among others.

Back to the mess deck we went. I had been designated to say breakfast grace and that made me proud. But I must say that was not an unusual event on *Dixie*'s mess deck. We all wished that our delayed brother (Hawkeye) could be with us and I mentioned that fact at grace. Our leader had a way of bringing us together so that we were happy and when the time came, we would fight with full confidence in his leadership. I forgot to inform the reader that *Dixie* always serves very good meals!

We were to the last man anxious to begin work on our boat, but we were required to wait until after colors at 0800 (8 a.m.). Chief had five new names added to his muster. After roll call the Sergeant and I got his permission to report to my doctor. My arm felt good but I knew better than to ignore an appointment with her. Anyway, seeing her won't take long, I thought.

Chapter 25

After a rather lengthy examination of my arm, the captain said, "Young man you have healed quickly, your arm looks very good. Sergeant, I wish for you to make sure this arm stays dry and out of the sun. I'd like to try using a lotion and eliminate the bandage, we'll see how that does. Also keep it clean!"

"Yes ma'am I will do that."

"I believe we are looking at about two weeks until the check ride if no further injury occurs. Do you like what you have just heard coxswain?"

"Do I! Will you marry me?"

Everyone got a huge laugh out of that. "Yes, if I were twenty years younger, but thank you for asking. Now you be careful and I'll see you day after tomorrow."

"I'll be here, thank you ma'am."

She gave the Sergeant the prescribed lotion for my arm and recommended morning and evening application. On our way to the boat deck we saw Hawkeye, in hand-cuffs as a disguise, being escorted up *Dixie*'s boarding ramp. We arrived at the quarterdeck about the same time he did. There was some paperwork which the Sergeant took care of, and then we took him to our – now his – quarters and had a loud reunion. All three of us rejoiced that we were back together again. I tried to answer his question, which was, "What's up?" but I passed it over to our leader because he knew more of the details about our new assignment. After the Sergeant's update on things he should know, we asked about his trip from Key West to Sasebo.

"I haven't slept in more than twenty hours and I am probably not good company at this time. Is it okay if I sleep before meeting my new teammates?"

"No problem, there is your made-up bed, sleep as long as you like! We will be on the boat deck if you should need us. I'm sure that you can find your way around on board *Dixie*."

"Oh yes, I can do that easy enough Captain."

The Sergeant and I reported to the chief that our lookout was safely on board sleeping after a long, tiring trip from Key West. He walked with us to the sailmaker's very large work table so he (Chief) could make sure the shelter and the insulated tiller were coming along as he wished. However, the sailmaker was glad I came because he was ready for my opinion about the tiller cover. "Is it too thick?" he asked.

I worked with it a few minutes and answered, "No it's not too thick but it will need strong lanyards to hold it in place."

"Thanks Captain. I had forgotten that."

"Have you begun making the spray curtains yet?"

"No but I'll start them in a day or so."

"Is it possible to use the shelter's uprights for the spray curtains also?"

"That's a good idea! At the moment I don't see why not. It sure would be more convenient to use them if I do it that way. Yes, I believe that it would."

"As you probably know, I'm having a periscope made. Could you make a canvas cover to protect it from the weather?"

"Consider it done, Captain!"

"I am very hesitant to ask but is the new shelter material as good as the old?"

"It is much better." Then he began to tell me in technical terms why it was better. The word "better" satisfied me well enough because all I wanted to know was would it keep me from freezing to death.

When he had finished, I thanked him, then we walked toward my boat. Sam was as happy as a robin in a chinaberry tree. He saw us coming and began talking before we got close enough to understand what was being said, because the boat shop was, at that particular time, very noisy with sailors at work. He repeated that an entirely new state-of-the-art AC/DC electrical system would be installed as soon as the boat got to an overhaul stage, and that it could and should be done. The electrician told me that after it had all been done, he would explain everything to me.

"AC/DC?"

"Yes, that's what he said!"

Sam helped me aboard to see our new engine. The chief and the Sergeant left to check on the new team members.

"What do you think Captain? We sure won't have to worry about a plowing bow with this one!"

"What about you Sam?"

"If we made it back with that other old worn-out engine, I believe that with this baby we will make it back home, especially with the best coxswain in the world and God at the tiller."

"Thank you Sam but you must understand that we are a team and that's why we beat the odds. Together we are the best! We think as one and that has many times saved our lives. Remember China?"

"Do I? On the coming mission, four of us have the benefit of experience, therefore we should prepare our boat with everything that we can think of that will get the job done and bring us safely home. I have been thinking about how we can get back on station (North Korea) without attracting any attention. Which reminds me, please tell whoever builds the new engine housing, to build a chart table for me. I prefer a fold-down top with storage underneath for my navigation things, which requires about two inches of depth under the top. Also tell the electrician that I need a two-way light over my chart table. One way white and the other way night vision.

"Now for something that could keep us from freezing to death. Whoever builds the engine cover must, with your experienced help, get as much engine heat as possible into the shelter! We need several appropriately placed electric fans to blow the heat from inside the engine housing through On/Off vents, appropriately located. On this mission heat from our engine must be near totally captured or we will not be able to remain on station very long. Sam, whatever ideas you have for that problem - use them!"

Noon mess call was sounded on the PA. "Time to eat, Sam. We will meet the other team members on the boat deck." We did and all headed for the pleasures of the mess deck.

Chapter 26

I wondered about Hawkeye but I guessed he was still sleeping. However, he could be walking right behind us and no one would know it because he doesn't talk much. I've often wondered about that. It wasn't fear, because in combat the man was outstandingly courageous.

This time going through the chow line I elected to carry my own food tray! The mess cook walked along with me just in case I couldn't do what I thought I could. Those who had been watching an injured sailor, for I don't know how many days, cheered after I had been successful. One shouted, "That-a-boy 'Boats!'" Some even applauded!

From time immemorial a Navy coxswain's nickname has been "Boats." Of course, I waved to all of my well-wishers. The Sergeant asked one of the new team members to give thanks for our food. He did with heart-felt words. During our meal, we all agreed that we were very pleased with the rapid pace of our boat's overhaul and praised the outstanding workmanship that was being done on it. Even though it had not been much talked about, we all knew that our peaceful comforts on *Dixie* were fast coming to an end. Soon we would once more be in the life-threatening frozen north. A place where it is necessary to constantly fight nature as well as man to stay alive. As always, waiting to go into harm's way became an inexplicably heavy burden on those who were destined to do it. If the Russians knew we were coming, they would use their entire Pacific Fleet to either capture us and use us as propaganda pawns, or kill us to satisfy their growing desire for revenge. Every day, I spent more and more of my time trying to, in my imagination, stay ahead of what I thought they might do so that I could do the opposite!

If we are in the middle of their free-floating mine field, they are not likely to pursue us with surface patrol craft, or for that matter with submarines. They probably will try to do us in with aircraft, therefore I should concentrate on aircraft defense using some kind of disguise, such as a lost Sampan, or radio equipment to call for carrier air support. After all, I will be trying to prevent mines from drifting into their carrier so keeping me on the job should be a very high priority for them.

I became so deep in thought that Sam had to shake my shoulder to tell me it was time to return to work.

"Is there something wrong, Captain?"

"No Sam, I was just thinking."

"About what?"

"It's just one of my mental adventures that I'm not quite ready to talk about yet! Sam, are there new and larger bilge pumps on our want list?"

"I don't think so."

"Please look into that matter and make sure that we have them."

"Aye, Captain.

"Also see to it that they are well lubricated and insulated. We often overlook things like that until our life depends on their proper operation. Is it possible to replace a thru-hull pump while the boat is in the water?"

"No Captain, there is too much displacement pressure."

"So if the pumps are gone we are gone. Let me think on that and I'll get back to you.

When we were back on the boat Sam piddled around the engine a while and said, "Look here Captain, I can do this one by myself!"

"Do what?"

"Make an emergency bilge pump."

"Show me!"

"I'll put a 'Y' valve on the engine cooling water intake pump, which will never freeze because it goes directly into a hot engine. Then alternate the engine's cooling water needs, which is a lot of water, between seawater and bilge water if necessary. It will pump out a large amount of water and will be quick and easy to manage if it is built as I plan to build it. Also it will not freeze or harm the engine in any way, and can be made very accessible to me from my seat box. We already have the monitoring equipment. For example, when the boat is pumped out and the engine begins to heat, simply reverse the 'Y' valve to put the engine back on seawater and keep it from overheating.

"Do you have any more problems like this, Captain?"

"Sam you have solved a problem that I have been agonizing over for many hours. I well know that if we should lose our two pumps, which were not even designed for the extreme climate we will soon be in, we are dead men. If you remember, I was worried about that possibility on our last mission."

"Yes Captain, I remember."

"I believe what you have demonstrated to me just now will be the answer I've been looking for and I am happier than I have been in a long time. Consequently, I must test my understanding. Are you telling me that if our engine is running you can, without the two small pumps, keep our bilge dry even with very significant flooding?"

"Bill you are my best friend and captain of our boat, which will soon be in desperate straits, and I could never mislead you. YES, that is exactly what I am telling you."

"Okay, I am with you on our emergency pump. If you should have any problem getting the parts you need send for me. I'll be gone for a while, the

Sergeant is putting me through rehab exercises so I can be re-certified as an assault coxswain. And since that has come up I can tell you now that I want my Navy crew to be with me in whatever boat is used for the certification. My doctor told me this morning that she will need to go along for the test ride, which will be done in about two week."

"Bill, let me say something to you before you leave."

"What is it, Sam?"

"I agreed to return only because you are still our coxswain. I was not sure you would keep your arm. But you can bet your last dollar that Hawkeye and I will be on that boat when you are tested for recertification. Keep up the rehab and you will get back your strength. Also remember I am praying for you every day."

"Thank you Sam."

Chapter 27

As I walked away I could not help but think what an amazing man Sam is. People like me are successful because of people like him. During my calisthenics routine the Sergeant raised the bar again and moved me from fast walking to slow running. I could tell that I was getting stronger every day and I was very thankful for the Sergeant's dedication to my full recovery. He was another amazing man to whom I without a doubt owed my life. I would think by now that anyone who has served with me knows that I am a very cautious captain and will accept nothing less than the best seamanship from my crew. When I met the Sergeant for rehab, I told him about the ingenious emergency bilge pump idea that Sam had come up with. "It will have an unbelievable pumping capacity made possible by our two hundred horsepower diesel engine."

He was very pleased to hear that and commented that beyond-our-control flooding had been a great concern for him more than a few times. Our two small bilge pumps were not capable of extracting one small wave if it came into our boat, even if they were operating perfectly. It was a wonder that during the blizzard they didn't freeze solid because they were not insulated, and then just one wave would have finished us off for sure.

"You know Sergeant, perhaps we should teach our new teammates how dangerous it is to be in a heavily loaded, twenty-eight foot, open-hull boat twenty-four hours a day, just inches from death on our best day."

"Good idea Captain, I will begin their education tonight. And may I say that I will need your help."

"You may!"

When we had finished my rehab program, I remarked that mess call had not yet sounded. He answered, "Well let's fast walk until it does."

"Sergeant, I'm not an Olympic contender! I just want to re-qualify as an assault boat coxswain."

We both had a little chuckle over that come-back. Mess call was sounded. We rallied with all of our teammates except Hawkeye and proceeded to the mess deck. I had already proven my ability to handle my mess tray on my own so the mess cook stepped aside and allowed me to do it. I was asked to give thanks for the food we were about to receive. After I had done that, the Sergeant announced, "There will be two hours of seamanship instruction on the boat deck this evening following chow (meal)."

He further emphasized, "We must learn and practice cross-training so that when a casualty occurs, and it will, someone will have been trained to immediately do the missing or disabled person's job. That kind of multiple training will greatly increase our chances for a successful mission."

"Sergeant, look who's in the chow line!"

He bolted up as soon as I directed his attention toward Hawkeye and rushed over to greet him. Both of them returned to our table and our new eight-man team ate together for the first time. Our leader began the long task of bringing our lookout up-to-date on everything. But eventually we all went to the boat deck to begin seamanship instruction.

The Sergeant asked me to speak first and I did. "The very first thing I want everyone here to learn is that there can be only one leader for any group for any purpose and that ain't me. I was chosen by the Sergeant to be his team's boat captain much in the same way each of you have been chosen! My track record was important but the most important thing was my pledge to our leader that I would without hesitation or question obey his order. That single pledge of obedience by each of us has kept this team alive and has made us internationally famous, especially with the help of Russian propaganda. Now remember this if you don't remember anything else I've said, this team is not a committee. It is, thank God, a dictatorship and the Sergeant is the dictator. I do absolutely nothing with our boat except by his order or permission. I will give an example: 'Captain take me to Vladivostok.' I immediately began giving my Navy crew navigation orders to get underway for Vladivostok. Plain and simple, there was, no delay and no questions!"

Next Hawkeye was called on. I remained with him to give him support. However, I soon learned that he didn't need me, so I moved and stood with the others. He revealed, "Being a good lookout is a matter of concentrating on what you are looking for, and since we are going into a sea of mines, the lookout's skill at spotting those mines is an awesome task. If you have ever seen one of those things explode, even in movies, then you will shudder to think what they could do to an MWB. Nothing I say here can equal what I will do tomorrow on our boat. I plan to teach each of you how to look. That's right, I'll teach you how to look in a way that will broaden your field of vision. I will also teach you about night sounds and how they travel over the water and reveal as much as or more than vision. Your ears and your eyes are the key elements to being a good lookout. And finally, while on lookout watch, never forget that you are keeping your teammates safe. It is my opinion that anyone who fails to do his very best while standing lookout watch, the most important job on any vessel day or night, should be shot and thrown over the side. I will be available to discuss this very important responsibility with any one at any time, except when I'm on lookout watch."

Sam was next, and he right away began to tell us about his new bilge pump invention, which would make our boat and all who are in her much safer. Then he began to explain other benefits of our big new diesel engine.

Chapter 28

"The most important thing is that we now have enough engine power to head upwind in an angry sea and reach any destination we choose, instead of being forced, because of insufficient power, to run downwind and end up wherever the storm might take us – even crashing ashore. A strong downwind without sufficient power is an invitation to 'pitchpole,' which means being driven by a powerful storm's wind over a rogue wave's top and into its deep trough, never to be seen again. That is every small boat sailor's worst nightmare. Another very good benefit," he said, "is our new canvas shelter, which will give us at least some protection from a sub-zero arctic wind. We all must understand that there can be no passengers in our boat, because the hard truth demands that every moment of our time be spent trying to keep her afloat while we are simultaneously doing the job we have been assigned. And remember this, if you fall into the freezing water, you will have little chance of being rescued alive because survival time is less than three minutes. Throughout our mission you will be constantly cold, hungry, and sleep-deprived but while you are suffering to save the lives of your teammates and many United States Navy sailors, your personal problems by comparison will seem trivial. You might be interested in knowing that many of our sailors have already been killed by those deadly floating mines, which are obviously increasing in number. If any of you have questions for me now or later, don't hesitate to ask."

Our leader, the Sergeant, was next. He began by informing us that "In addition to mine source detection and destruction we have also been ordered to intercept any Sampan we sight more than ten nautical miles offshore and destroy it."

I didn't like that order because there are often families in those boats. However, I later learned that there had been attempts to bring Sampans loaded with mines into the paths of our Navy's ships. Anyway a legitimate

Sampan fisherman seldom goes more than about five nautical miles offshore. I returned my attention to our leader.

"You must remember that any Sampan could be a death trap. Some will hold up children to cause us to hesitate, and if that ploy works they shoot as soon as we are in range of their (.31 caliber) antique rifles.

"Well gentlemen, we have gotten off to a good start. We will meet again tomorrow evening, at the same time and place. We have seven days left on board *Dixie* so let's make the best of the time we have left. I do not want to get on station and discover that we have forgotten something that's important to our mission. That's all for now!"

The Sergeant and I remained in place for a private talk. "Sergeant I have thought of something that we should consider."

"What's that captain?"

"A urinal!"

"I have thought of that also but there's not enough room on the boat, even though it is badly needed."

"Oh yes there is!"

"Where for heaven's sake?"

"We will have the shipfitters build one from my design."

"Hey, that is great! Show it to me!"

"Follow me to our boat."

On our arrival, I showed him where. "A two-inch diameter housing for a drain line could be permanently attached to the outside hull from the gunwale (railing) to the water line. From there it's simple, just attach some kind of apparatus (funnel) on the business end of a one inch by five foot drain line, and shove it into the drain housing. That apparatus should be pointed aft about sixty degrees to create suction on the drain line from our forward movement. Now that's a very crude explanation but any improvement of my plan will certainly be most welcomed."

"Captain, you have solved, as usual, a big problem in a very simple way. I'll take it from here and see that it is done."

"Okay, call me if my further involvement becomes necessary."

"It's getting close to taps, Captain, so we had better return to our quarters."

The following morning after roll call, the chief asked to meet with the Sergeant and me to evaluate progress on our boat's overhaul. To our surprise, progress had moved considerably beyond our expectation. Chief was very confident that our boat's refurbishment would be completed on schedule as planned. The Sergeant brought up the urinal idea and began telling the chief what I had said about it. He liked my idea and asked for me to show him exactly what I wanted. I did and he became even more interested. After observing once more both inside and outside of our boat, the chief said, "Let me work on this a while, I'll get back with you after noon mess call."

"Thanks Chief!" The Sergeant and I went to our respective work stations. He (Sergeant) had told me earlier about how well the shelter was coming along and that several extra things that we didn't know about had been added. After the Sergeant had told the sailmaker how ferocious the blizzard had been, he double-strengthened everything. They had established a professional friendship, so the sailmaker had taken a personal interest in the quality and strength of our shelter. I climbed aboard our boat before Sam saw me do it, and I got reprimanded.

"Don't scare me like that, Captain, I'm right here. Please call me before attempting that again. You almost died on me once and I will not permit that a second time."

"Okay Sam, I promise I will call you before coming aboard alone anymore."

"Good. Come over here, I have something to show you. See my two nice, dry battery boxes, and how they have been built in a very serviceable place."

They were larger than the others because Sam had told the electrician he would need more cold startup power. The electrician had begun work on our boat. However, it would take two more days to finish.

Chapter 29

"New and larger fuel tanks will be installed tomorrow."

"How much larger?"

"There will be two newly designed seventy-five gallon tanks, giving us a total of fifty gallons more capacity! They will be installed with my careful attention to the boat's righting moment (weight distribution) including the weight of our new engine."

"Have you made your moment marks?"

"Yes, I personally made sure that was done."

"That sounds very good to me Sam!"

"And another thing Captain, the new tanks have inside dividers to keep the heavy fuel load as stable as possible during stormy weather."

"Now that, I'm very glad to hear. I have fussed about that dangerous problem more times than I care to remember. More goodies from the admiral I suppose!"

"That is exactly what I thought!" Sam agreed with a laugh.

After the good news about the fuel tanks, I asked him to supervise the installation of a urinal I had designed and which I believed could be done.

"A what!"

"A urinal!"

"Bill don't kid me now."

"I'm not kidding! Come over here and I'll show you where and how."

We agreed on a location but decided to leave the how-to problem for the shipfitters to solve. "Captain a urinal will be a wonderful addition for our comfort and it's so simple. Why didn't someone think of that before?"

"Because no one has ever done with a MWB what we have done."

"Now, what about a 'Potty,' Captain?"

"I believe that between you and the chief that problem also can be worked out. The admiral has one in his barge, need I say more?"

"Consider it done, Captain."

"Remind the chief that below-zero weather is a very risky and a very uncomfortable place to expose oneself – particularly outside. Also remind him that we will be ten miles or more offshore in the middle of a mine field protecting the fleet, which includes *Dixie*'s destroyers, from those floating mines."

Mess call was sounded loud and clear. "Let's go Sam, it's chow time."

We met the others and proceeded to the mess deck. Each one had a story to tell about their morning activities, that often with cheerful enthusiasm was interrupted by another enthusiast. Everyone seemed to be getting more and more excited as our departure time to the frozen north was drawing near. Our lookout was surprisingly the most aggressive talker of all while telling about his training the new Rangers how to be lookouts. We all laughed, talked, and ate our fill, because each of us knew we were very close to K-ration time again.

After we had returned to the boat deck I mentioned to the Sergeant that I needed Chinese, Russian, North Korean and Japanese flags the same size as our boat's American flag. "Also it would be very helpful," I added, "if we could find a way to make our boat appear, from a distance, to be a Sampan dropping floating mines into the Sea of Japan's south-flowing currents."

"I'll take care of that as soon as I get back, Captain."

"By the way, how is our shelter coming along?"

"You are going to like what has been done so far, I guarantee it. Right now the boat is covered with workers so we must wait before we can test it."

"As you already know, Sergeant, I'll need to wear my survival suit while I'm at the tiller, which will be most of my time. But my face will still need to be protected from frostbite. Now that I'm thinking about it, I recommend that you request face protection for every man because a man with facial frost bite is finished as a lookout. Therefore, without face protection, we very much increase our chances of colliding with a mine. And it goes without saying that this mission is all about seeing floating mines, searching for their source, and then destroying as many as possible. Several Navy minesweepers have already been lost while sweeping off the coast of South Korea. But our primary search area will be in the area of Vladivostok. Russians claim that everything up to one hundred miles off their coast is Soviet territory and they will take appropriate action against all who ignore that claim. Since we will be the only eyes our fleet will have in that area, we should have everything we need to do the job, and we are the only ones who know from our experience what we need. I recall that we had on our want list eight harnesses with ten-foot safety lines (rope) attached. That may sound unnecessary to some people, but to us it is a matter of life or death! However, no one has yet refused to do anything we asked so perhaps my concern is a bit premature."

"I don't think so, Captain. You are much like I am, it's down-right frightening to think about being in the most forbidding place on Earth and discover that something important for survival has been forgotten. God helps those who help themselves and that is what we are doing. Making sure, as best we can, that we do not forget anything. Captain, I want us to meet today for rehab training at three-thirty. I believe you are ready for that much-extended time."

"Well if I am not, I had better soon be ready because there is little time left on *Dixie*."

"You'll be ready!"

We separated with each saying to the other, "Three-thirty it is!"

Sam saw me coming and prepared to help me aboard. The electrician had come earlier than expected and that pleased me because of my need to talk with him about communication between the lookout, Sam, and me. After I was on board, he began telling me what had already been done and what he planned to do. There was a verbal pause, so I jumped right in and began telling him about the critical need for communication between the lookout, Sam, and me.

"Show me where you want the headsets."

I asked Sam to handle that because he would be there all day to answer questions. Also he knew more about it than I did.

Chapter 30

It seemed the overhaul of our boat was rapidly moving toward completion day. Many small things had already been made ready for our next mission and even some large things had also been finished. Watching a crippled, war-weary assault boat being brought back from near junk to an almost new fighting machine was for me like magic. However, my boat could not be used for the check ride tests because it would not be totally ready in time for that all-important event. Assuming my doctor would tell me tomorrow that she believed I was physically ready for re-evaluation, I could be evaluated the following day. With so much effort being made to get our team back on station (Vladivostok) as soon as possible, it became obvious that the floating mine situation was causing big problems for the U.S. Navy.

"I'm certain the admiral is not willing to wait until my boat's restoration is complete just so it can be used for my recertification. My guess is that he will use one of his destroyer escorts to speed up the process, which is always a series of time-consuming navigation and seamanship exercises that are usually followed by dozens of irrelevant questions. All of which would come from Liberty Boat coxswains who would probably get lost in San Diego harbor."

It might be well if I explain to the reader why our Navy chose a twenty-eight foot wooden-hull boat for the floating mine mission instead of a large, well-armed, steel-hull ship. The answer is **stealth**, which is a must if one wishes to prevent being blown out of the water by Russian submarines and/or aircraft. A wooden-hull boat like mine that has been radar sanitized has total stealth capability. Notwithstanding the fact that four of our assault team have had previous covert experiences in Russia, which are very valuable to help carry out our newly assigned mission. And that mission is to locate the source of those deadly floating mines and then determine the course and speed of the sea currents they are using to float them toward our fleet. It may be necessary at times for the coxswain to float along with those unpredictable mines to get much-needed information for our fleet course plotters. A large, steel-hull vessel can't do that, especially in Russian territory, because the large intruder would be instantly detected by radar, and sunk in a heartbeat. At that time the USA was trying every way possible to prevent a declared war with Russia and China, even though they were deeply involved in the Korean War and everyone knew it. So we used the big stuff only in Korea.

It is very important to concentrate on a selected mine to obtain accurate plotting data for our Navy, and permit me to say that concentration in a mine field is a good way to stay alive. I truly believe that no other vessel in the U.S. Navy can do better what we are about to attempt! And I also believe that no one else would have a Chinaman's chance of pulling this one off except my crew (assault team).

Now I'll share a little humor from our team's previous experiences with floating mines. To relieve the pressure, we assigned names to mines that drifted too close. For example, "Betty Grable is getting too close to me!", "Poison Charley is crossing our bow," and so it went, hour after hour. We disguised our boat and remained as safe as possible from surface craft because no one would dare pursue fools like us into the middle of a floating mine field. The above is part of what we did while operating from Nando Island (previous mission) and will continue to do on our coming mission. I hope the reader will now understand why a small wooden boat, like ours, with an experienced crew was chosen over a battleship, which is far too big to do the job and has zero stealth. The bottom line in our business is to be undetected as long as possible.

Time to meet with the Sergeant for rehab, I thought. He probably has already anticipated from the daily rehab records that tomorrow would likely be the day the captain will cut me free. If she doesn't, after what he had put me through today and I had still remained alive, I would not even guess when she would consider me physically fit enough to be reevaluated for assault coxswain duty. However, I'm sure the admiral would have his say about any further delay, and admirals seldom lose an argument.

Mess call was music to my ears and when the others arrived we lost no time getting to the mess deck. As always, everyone was filled with good cheer and good reports about our boat. The Sergeant suddenly reported to us that the chief had told him our boat would have her final inspection the next day. That announcement filled my mind with caution, because every sailor knows that after a vessel has been overhauled and launched, sea trials are always conducted. During that time serious errors are often found and corrected. After we had finished eating, I recommended to our leader that everyone return to our boat and conduct our own inspection. "Believe me when I say it's a matter of life or death. Think how nice it would be to present the chief with our own inspection report."

Everyone agreed so there was no further need for seamanship training after the evening meal. There were eight people trying to insure their lives. "It has been my experience that when hurry-up is involved in anything there will be neglect and/or mistakes. It is very obvious that the admiral, or someone even higher up pushing him, is not willing to wait one more day for a sea trial of our boat. Therefore, we will have a dry sea trial – that is all we can do. Every system will be thoroughly inspected, and every nut, bolt and screw will be checked for tightness. There will be boat shop workers there to assist us with our concerns and they will, I'm sure, joyfully welcome our help."

Chapter 31

The chief saw us as we entered the shop and walked over to thank us for coming to help because he was running out of time. Suddenly there were about twenty men on and about our boat. The Sergeant cautioned me to be careful and remember that my physical exam was scheduled for tomorrow. I promised that I would and asked him to help me examine the rudder post. It had become somewhat difficult to move during the Sea of Japan crossing and that would be totally unacceptable in a sea of mines where fine-tuning a course would be continuously needed. We pulled up the very large connecting rod (one inch by 38 inches) enough to see that it had not been lubricated since

Sam and I had done it at Nando island. The chief was close by, therefore I asked him to fetch some heavy duty grease so I could free up my tiller. Soon he returned with the grease and helped lubricate the large connecting rod. I like for my tiller to be resistance-free so that it will respond to inches, if that's what I need.

"Hawkeye, when you have a moment to spare, I need you."

He came to me immediately and asked, "What do you need, Captain?"

"Has the sea-anchor cleat been sufficiently back-plated?"

"I'll have a look. There is no back plate on it at all."

"Please see to it that one is properly installed."

"Aye, aye sir."

And so it went throughout the night. However, at three a.m. the Sergeant insisted that I go to our quarters and sleep until reveille. I bathed well for my appointment with the captain and fell asleep five minutes after I lay down. But it seemed that only five minutes had passed before reveille blasted away again. Before I was fully dressed the team came into our quarters and began telling me how much progress had been made on our boat. They were all very excited and believed that by the time I had finished with my medical exam everything would be completed.

"That sounds great, I'm looking forward to ending my need for medical appointments today but I have no idea how long this one will take. That U.S. Navy doctor prevented my right arm from being amputated so I will be patient with her for as long as it takes."

After breakfast we assembled on the boat deck for roll call, where one after another boat shop worker told us how special they had made our boat. The chief already knew that the Sergeant and I would be going to visit the doctor immediately after roll call. When we arrived at the medical facility (sickbay) there were six assistants with her and she had added the Sergeant, which made it seven.

"Good morning coxswain, I'm sure you know why you are here!"

"Yes ma'am I do."

"All right, after you have removed your clothes, we will begin. My assistant will help you, please go with him to the dressing room."

He instructed me to put on a gown which was at the extreme edge of my modesty, but as I have previously written, I would be patient and cooperative. We returned and he asked me to sit on a medical table which had been prepared for the examination. Then the captain revealed that my physical examination would be a teaching event for Navy corpsmen (nurses).

"Is that all right with you coxswain?"

"Yes ma'am it is."

Once again people were all over me doing this or that. The Sergeant held all of my medical and rehab records; consequently, he remained busy answering their questions. Including the teaching time using my prone body as a prop, the physical examination took the better part of three hours. While I was dressing, the Sergeant pulled back the curtain on the dressing room entrance and said, "Congratulations Captain, every thing is fine and dandy. The doctor will tell you more completely when you are outside. Come on, they're waiting."

On our return the Captain said, "Young man, I have found you to be among the healthiest people your age that I have ever examined."

" Does that mean...."

"Yes, that means so far as your health is concerned, I will recommend your return to regular duty as of this day. But there are others who will have their say about your recertification as an assault coxswain. However, we will find out about that tomorrow, won't we?"

"Yes ma'am. Now it is time for me to keep my promise." I put BOTH of my arms around my doctor and gave her a big well-deserved hug. Everyone present smiled and applauded.

"Thank you young man, it's not every day that a lady gets hugged by a sailor."

"And don't forget my test ride tomorrow."

"Ma'am I'll refuse to go unless you are there!"

She laughed and said, "Goodbye and have a great day!"

I could not stop thinking about the fact that she was the only one who, at the end of all medical possibilities, stubbornly refused to amputate my right arm.

"What about that, Sergeant?"

"All of the medical literature agrees that you should not be swinging that arm as you are right now," he answered. "God without a doubt was involved in this miracle."

"I'm glad to hear you say that, Sergeant, because I know that to be a fact. Just think of all the special people He provided to care for me in my time of great need. People like you and our teammates! Do you think He will go along with us a second time?"

"Indeed I do Captain!"

"I'm all fired up and ready to take the recertification test ride right now if you wish. But I will not let those Liberty Boat coxswains play around with my time, if they want a boat ride they will get one. I will promise you that before I'm through they will be begging to certify me. Do you remember our ten-knot launch from *Perkins*?"

"Do I remember it? … I'll never forget that boat ride."

Chapter 32

"Well, let's allow them to see those deadly propellers with a lee-side view from a ten-knot launch and that will forever imprint danger on their minds! And I want the pick-up to be ten-knots, also on the ship's lee side. I'm the only person in the United States Navy who has ever done that exercise and I can still do it! Anyway, working on the lee is much more exciting than on windward."

"How about that lady passenger you will have on board?"

"Oh!" I had momentarily forgotten that. "Please explain to her what I am planning to do and that I will not put us in any danger. Also I promise to make the ride as smooth as possible. Make our arrangements as if the queen of Siam's personal barge were being used for this test ride."

My leader was not only an outstanding military man, he was a master at managing details. All arrangements had been made and after breakfast my doctor, Sam, the Sergeant, Hawkeye, and I left the *Dixie* and boarded a destroyer escort moored close by. (An escort is considerably smaller than a regular destroyer.) They had been waiting for us so in less than an hour we got underway.

About ten nautical miles offshore, the exercise began. We were all standing beside the MWB which would be used for the recertification test when a chief boatswain's mate looked at me and said, "It's your boat, coxswain."

"Thank you Chief! Okay mates. Let's have a prelaunch check before she's hoisted out of the cradle."

Sam started the engine and checked the fuel and other essentials. Hawkeye and I checked all of the inside hull and required equipment. We had already looked over the outside hardware. He found an absolutely necessary lanyard missing from the forward fall and I found the bilge plug was missing.

I reported those missing items to the chief. He walked over to me and said, while handing the missing items to me, "Well done coxswain."

I engaged the tiller and asked the Sergeant to observe as I moved it full port and full starboard. "Now I'm ready to board my passengers. Chief, please help the doctor aboard."

After everyone was onboard and seated, I ordered life jackets for everyone.

"I'm ready to be lowered to the rail, chief."

Just as we were about to be lifted off the cradle, the ship's captain was seen approaching us. He came close to me and asked, "Are you the coxswain who crossed the Sea of Japan in a whale boat?"

"Sergeant, will you handle that question for me please?"

When he had answered the question the captain turned back to me and asked, "Are you sure you want ten knots for launch and pick-up, both on my lee side?"

"Yes sir I am sure because that's necessary in the frozen north to deceive submarines, and my doctor and I must know if I can still do it."

"All right but if I sense any trouble, I'll stop both of my engines."

He expressed good luck as he walked away.

"Thank you Captain! Lower me to the rail if you please, Chief."

He called to his launch crew, "Lower her to the rail!"

While stopped at that level, I ordered Hawkeye to attach the sea painter. "Let me hear a station report!" I shouted.

"Ready in the bow, ready in the engine space, are you coming aboard Chief?"

"Yes I am!"

"Get him in a life jacket, Sergeant."

The captain turned on the green light from the bridge deck, indicating that everything was on go. "Okay Sam the ship is at ten knots. Are you ready?"

"Let's do it, Captain. Lower her level and easy, Chief, and try not to let a swell run out from under us and leave the boat hanging with no flotation."

The launch crew really knew their business, and Sam was perfectly matching the ship's forward speed.

"Stand by to release the aft fall." I had trained the Sergeant to do that. "Execute! Hoist the aft fall!" I shouted. "Stand by to release the forward fall, execute! Hoist the forward fall!" I shouted.

And now for the most dreaded one, although she had been running true because she was not heavily burdened with cargo. "Stand by to release the sea painter. Execute."

I instantly felt she was free on her own very sufficient power, and with excellent control I eased her clear of the escort's steel hull and deadly propellers. Sam looked at me and smiled. I knew we had scored big.

The chief said, "That is the best I've ever seen that done, and as a matter of fact I've never seen it done over three knots."

My passengers and crew swore that not a drop of water got into the boat. That day we had two- to three-foot swells and that helped a lot. The mother ship's captain sent "unbelievable" on his blinker light, I answered with "thank you sir." His next message instructed me to heave-to about two thousand yards (one nautical mile) off his port side. I answered, "understood." When I was about one mile off his port beam, I signaled with my blinker that I had hove-to as ordered. We communicated back and forth with our lights for about ten minutes, after which I was ordered to read his semaphore. I immediately realized his signalman was sending too fast for me to interpret, and with my flags I requested that he slow down. He did and the problem was solved. After ten minutes or more of semaphore communication we moved to international flag signals, which took almost an hour because they required moving my boat's location according to my interpretation of the masthead flags. Finally I received, by blinker light, "well done coxswain bring my boat home." With my blinker I answered, "aye aye sir."

I informed my crew about the good news and added, "Okay men, let's show them how a MWB can be picked up at ten knots!" Sam motored up to a peaceful cruising speed and I steered to enhance that peace. The Sergeant asked the captain (doctor) to sit by me in the cockpit so that he could take our picture.

Chapter 33

"Oh! I didn't expect to have my picture taken."

The Sergeant extended his hand and said, "Please ma'am!"

She agreed and after several snapshots I let her steer the boat, which really thrilled her. She returned to her seat and I began preparing to be picked up.

"Sergeant, please ask Hawkeye to report to the cockpit."

There I told Sam, the Sergeant and Hawkeye my plan for the ten knot pick-up. "I will approach the port propeller guard about twenty degrees off center. Sam, do we have enough catch-up power if our target is doing ten knots?"

"Yes, I can gain on her two to three knots per hour."

"Good, that will assist me in knowing when to make my turn toward her stern. I want the doctor to see the propeller in action, so I'll move in close to the port propeller guard, and then just behind the boat davits (hoist), I'll line up parallel with the ship's hull and keep the boat's clearance from the hull about five foot while Sam puts Hawkeye in a position where he can reach out and take the sea painter with his hand, then calmly secure it on the Samson post. I don't want to see any messenger lines thrown to him, and I do not want any one hurrying about as if they were not sure what to do during the pick-up. But I do want all of us to show absolute confidence. If I determine there is a problem, I'll go around and we'll try again. Are there any questions? (silence) Good! Sergeant, you and the chief will be responsible for the doctor."

"You can count on that Captain."

"All right mates, let's go show 'em how it's done."

When the time was right, I lined up the boat on my chosen course to the port propeller, which was kicking up excitingly beautiful spray and sea foam. After only a brief pause to watch, Sam moved us forward along the hull just as we had planned. Hawkeye reached out with his hand to a stunned sailor, who was holding a messenger line ready to throw to a bowman who needed help. Instead, Hawkeye gently took the sea painter from the sailor's extended hand. At the moment he had calmly secured it on the Samson post, I knew we were safe, and consequently pulled the drain plug while we were being hoisted, but there was no water to drain. The decks were lined with sailors who had come to see either an accident or a coxswain who had failed to achieve his goal. The falls (hoisting blocks) were attached in their must-get-it-right sequence with equal calm, and as we were being hoisted up to the rail the sight-seers went wild. They were applauding, cheering, and taking pictures. Many of them held the boat steady and in tight against the hull while the Sergeant and the chief helped the doctor onto the mother ship's main deck. The captain was there waiting to congratulate all of us and had a boatswain's mate with him who piped the doctor aboard. It was a special, warm feeling for me to see the woman who had kept me from going through the remaining years of my life as a cripple being treated with the respect her rank deserved. She was graciously escorted to the wardroom for a high-ranking officer's guest lunch. The boat crew had a designated table on the mess deck and the meal was delicious. The backbone of the Navy (non-coms) had really shown their colors that day!

Somewhere close to two p.m. I was summoned to the wardroom were I was told by my doctor that my recertification as an assault coxswain had been approved by the committee.

"Thank you ma'am," I responded.

The chief had been asked to appear before the committee and give his opinion concerning my recertification. But he arrived late due to a problem he could not walk away from. But he still wanted to speak to the committee.

"Permission granted, Chief," the captain answered.

"Thank you Captain. I have been a boatswain in the United States Navy for twenty-four years and this man," pointing to me, "is the only natural coxswain I have ever seen. What he can do with ease in a boat is totally impossible for any other seaman I've ever seen or heard of. So I strongly recommend his immediate recertification."

"Thank you Chief. You might like to know that the committee voted to re-certify him minutes before you arrived."

"Well sir, I'm glad to hear that."

The chief walked over to me, took my hand, and wished me well. I thanked him for his comments and wished him well also. I shook hands and thanked each committee person in the wardroom before I left. My teammates were waiting for me on the quarterdeck.

"Well, we did it, I'm recertified! Now we are ready to go to war again. Speaking of war, there are so many things left to do before we shove off tomorrow.

"Is that official?"

"No, but I believe they have only been waiting for me to get well and be recertified. Since I have done both, I'm sure it won't be long. When we return to *Dixie*, maybe we can find out something about what they are planning for us."

The trip back to the destroyer escort's mooring berth took about an hour. The five Dixie people gathered on DE's quarterdeck, thanked everyone concerned for their help, and asked the quarterdeck watch officer for his permission to go ashore. Permission was granted after we had each saluted the watch officer and our flag. Fifteen minutes later we reversed the procedure to board *Dixie*. On her quarterdeck we thanked the captain for everything she had done for us and promised to see her before we left. But little did we know, as we watched that wonderful human being walk away, that none of us would ever see her again.

In the middle of the night, leaving no evidence that we had ever been on board *Dixie*, we were, with everything we owned, transferred to the destroyer escort (DE) that we had just left.

Chapter 34

It didn't take much gray matter to figure that this was the vessel that would soon take us to the floating mine zone. Our boat was transferred from *Dixie* to the DE about four a.m. The MWB which I had used for my recertification requirements was strapped onto the DE's cradle so that the hoisting davits could be used to lift our boat. We got underway about dawn. Our team immediately got busy making our boat combat-ready, because absolutely nothing of the kind had been done on the *Dixie*! There were many needed things put into our boat at the last minute and equally as much was put on the DE's boat deck. However, with no certain knowledge of what had been left for us to do, we had no idea how quickly we could become battle-ready. Our team, needless to say, had a very busy time trying to prepare our boat for the unknown perils of a floating mine field. The boat shop workers had put extra things into the boat and on the DE's boat deck because they did not have enough time to complete the renovation of our boat, and that put us at a very high level of danger.

The Sergeant spent most of his time in the radio shack fine-tuning communication with his commanding officer, and I spent most of mine at the navigation chart table coordinating the DE's charts with those that the *Dixie*'s navigator had given to me.

I can now calculate a near perfect position which is a great improvement over the last mission, I thought. They even gave me a new sextant!

The DE's navigator invited me to join him in taking a noon sight (Sun's location relative to Earth). My margin of error that day was three miles and his was two miles. Not bad when one considers that we were approximately twelve thousand miles west of the prime meridian. However, the reader should be reminded that the DE's bridge deck is a much more stable sighting platform than a perpetually moving MWB. Under most circumstances, I believe the margin of error from my MWB platform would average five miles. When a navigator reports his position he should always give an estimate of his usual margin of error. During the long day that I worked with the chief, I learned that the DE would be our mother ship.

Wow! That is great news, I thought. Even though she is small, there will be hot food and a hot shower.

He asked me about the grounded WW II LST (landing ship tank), in South Korea, which I had referred to. He had not been able to locate it. We both had a thorough look on the large scale area chart and had no luck.

"Let's try a smaller scale map," I suggested.

"Smaller, or larger?" he asked.

"Smaller," I answered.

With the smaller scale we quickly found what we were looking for and I marked my chart for future reference. Mess call was sounded on the PA and I was ready for it. The navigator was a chief quartermaster, and chief petty officers have their own private dinning facilities, so he went one way and I went another, leading to the mess deck. DEs are small, slow ships, therefore they require a small crew of only about two hundred men. I met the team at the stairway leading down to the mess deck. Everyone was exploding to tell me about the huge amount of extra supplies the boat shop workers had left for us. "And we all appreciate that but the chief boatswain's mate is having trouble trying to locate a place out of the weather to stow them." I responded.
The Sergeant spoke up and said that the chief gunner's mate had assigned a gun locker for him and it had been very helpful.

He continued, "Some of you might have heard that this ship will be our mother ship for I don't know how long. I doubt if we will ever be more than two hundred miles from her."

Sam caught my eye and said, "Captain, I have a problem."

"What is it, Sam?"

"I must find a way to conduct a sea trial with our new engine. There are things that cannot be done without the boat being in the water and running with full throttle for a while."

"I understand that Sam but your problem must be presented to our leader."

"I'm sorry I forgot, but this problem is very important if I am to prevent engine failure in the middle of a mine field."

The Sergeant spoke up and said, "You bet it's important, Sam. I'll discuss this with the ship's captain immediately after our meal. I have one question though. Since our captain and I are very tied up with communication and navigation issues, do you suppose that the chief who was with us on the recertification trip would coxswain for you?"

"Great idea, Sergeant. I'll ask and let you know."

"There is one thing further, Sam."

"What would that be, Sergeant?"

"Launch and pick-up will be restricted to three knots."

Everybody roared with laughter when he said that, including Sam. We ate a quick lunch because we were all very pressed for time.

"Sam, I'll talk to the chief before you get underway if you wish."

"Thanks Captain. That will be reassuring for both of us (Sam and Chief) because I know exactly how you think and what you are about to do."

"I'll be in the chart house, just send for me anytime."

He sent for me in about an hour and I followed the messenger to our boat. Sam and the chief met me and shared the good news that the chief had agreed to coxswain our boat for the sea trial. After thanking him for the important favor he was doing for us, I ask if I could help in any way. He responded that his men had already been drilled on the launch and are waiting for the word to go.

"Sounds good to me. What launch and recovery speed do you plan to use?"

He looked at me with a smile and answered, "Three knots. The Sergeant and I have already cleared that with the bridge watch officer."

Chapter 35

"Well the sea looks real good for a high-speed run, good luck!"

"Thank you coxswain!"

I remained where I was to watch the launch, which was well-executed. Sam did exactly what I knew he would, as soon as he had cleared the DE's hull he slowly opened the throttle to hull speed. It was a thrilling sight to watch.

I'll wager she will go hull speed with a three-quarter throttle, I thought. She has good speed and appears to plane very well. This pleased me so much I could hardly walk away. But I returned to my chart work and kept looking out the porthole from time to time. When they returned I was standing on the boat deck waiting. The chief approached me and asked where I got that engine. I told him that I had ruined the one I used crossing the Sea of Japan because I ran it full throttle for thirty-seven hours, and when I requested a new, larger one, that is what I got.

"Do you know where that large engine is used?"

"I do, in an admiral's barge."

Suddenly, the Sergeant came rushing toward us shouting, "Mines have been sighted near us. Get the boat down to the railing, quick."

The ship was up to ten knots sooner than I thought was possible.

"Can we go after them, Sam?"

"Indeed we can, Captain."

The Rangers quickly returned with their weapons and prepared to board our boat. The Sergeant continued to the bridge deck but quickly returned and got everyone on board with their weapons. After we had been lowered onto the water, I delayed having the sea painter toggle pulled until the captain turned on the green light.

Hawkeye looked at me and shouted, "Green for go, Captain."

"Pull the pin NOW!" I ordered. There was very little feeling of change from being towed because Sam had so much power that we moved a bit faster than the ship, so I easily steered away from it and called for a course.

The Sergeant answered, "Northeast for now."

I pulled my compass periscope into place and lined her up on the course I had been given and reported, "On northeast course Sergeant."

He was on the radio talking with the captain of our mother ship when he looked at me and said, "Steer course 055."

"Steering course 055 Sergeant."

"Okay gentlemen it looks as if we have a fight coming up. The bandit is approximately ten miles away but we will finish him off and then look for mines."

Hawkeye already had on his headphones, Sam was putting his on and so was I. After we had all put them on we had excellent communication. I cautioned Hawkeye to use his eye protection because at the speed we were making the cold air could harm those valuable eyes! Sam told me that we were making hull speed, which is twelve knots and the throttle was only three-quarters open.

The reader might like to know the definition of "hull speed." Well here goes. Every boat's hull design determines its hull speed. Adding more engine power will not push a boat beyond its hull design, but it can get the person who tries to do it killed. A savvy boatman will know when his boat has reached its hull speed, because added engine power will not produce more speed. That is why Sam did not open his throttle fully. He knew that our boat had reached her optimum speed at three-quarters throttle. However, there are some conditions that might not fit the hull speed formula, such as surfing, or riding a fast current, but that is not engine power alone, that is flooding (a flood being an incoming current).

The sea conditions that day were perfect for fighting and fast running. The Sergeant had his Rangers standing at the ready, to do whatever he ordered. He and Hawkeye continually searched our line of attack for mines or a Sampan. My steering the boat a little port and equally starboard expanded the search pattern for our lookouts.

"How much time on the engine, Sam?" the Sergeant asked, using his headphones.

Sam answered back on his phone, "Thirty minutes."

The Sergeant's message to me was "Stand by for a fight coxswain!"

"Understood loud and clear, Sergeant."

In no more than ten minutes, Hawkeye reported an unidentified object two points off our port bow. "Correction, Sampan two points off our port bow!"

The Sergeant prepared his Rangers to shoot from the bow – two kneeling and two standing. He ordered me to continue full speed directly to the target's broadside, so his people could make broadside shooting patterns for better effect. And since only our bow would be exposed, we would have less chance of being hit by return fire.

The Sampan obviously did not have an outboard motor, which meant that it was not on any radar screen. At about two thousand yards I heard the rifles lock and load and I knew the shooting would start any time.

Suddenly, I heard the Sergeant shout, "They are shooting at us – FIRE!"

In seconds there was a humongous explosion. The Sampan had obviously not completed unloading its mines.

"There is another one dead ahead," Hawkeye reported.

"Go get him coxswain, use the same attack plan."

I motored through the remains of the first Sampan and bore down on number two. I caught it in five minutes and the exact same thing happened!

There are two certain things about the two Sampans that we had destroyed: 1) they were delivering mines for the Soviets in the path of our fleet and 2) they had not unloaded all of their mines. And on top of all that they were shooting at us in international waters. In fact, they shot at us first. We looked all over for a number three with no luck. The Sergeant called our mother ship and reported what we had been doing. They agreed on a meeting course so we could return on board before dark. That worked out well and we were back on board before dark.

Chapter 36

After our first fight everyone on the DE went bananas! It was the first combat of the war for that slow, very poorly armed little vessel. And there was little doubt that every crew member's hometown newspaper would publish an article about what their ship had done on that day, because the floating mine problem was big news for the Navy at that stage of the war. However, our team's fighting style, hit and run, could easily identify us, and the Russians would get all fired up again to stop the embarrassment that our troublesome little boat had been causing. There were no survivors to tell the tale, but our enemy could get all of the information they needed from any American newspaper.

At the evening meal it was like Christmas for the DE's crew. One thing that was said still stands out in my memory – "Maybe now they will let us fight instead of deliver mail."

It had been a somewhat strange fate that the DE and our team had come together. The plan was for her to ferry us about halfway between Sasebo and Vladivostok but circumstances had changed that intention because the assault team and the DE had discovered, by working together, a very effective way to fight the Navy's worst problem – floating mines. That miracle was what high command had been praying for and now that a solution had fallen into their laps, they would not likely change it.

Our mother ship's captain decided to heave-to (stop) for the night, use whatever outside lights he deemed necessary, and post mine lookouts around the decks.

I'm sure Russian submarines will not harm us, I thought, because we are in international waters. But they will watch everything that we do.

It's interesting to think how totally illegal it is to endanger peaceful navigation in the ocean/sea far beyond the war zone, yet that is exactly what happens with free-floating weapons. And the Soviets are doing that daily and blaming it on a country (North Korea) that cannot manufacture a teapot on its own, let alone a floating mine

Now that the DE crew had an interest in helping to finish our boat, we had more help than we could use. We worked throughout the night. A shipfitter completed installing the urinal, Sam's bilge pump, and numerous other things. Others put up the shelter and the sea spray curtains. The ship's medic checked our medical supplies and added some things which he discussed with the Sergeant. When he was satisfied, he put the supplies in the locker which had been made for them and stuck a large red cross on the closed door. The engineer from the DE's MWB began helping Sam right after our evening meal. Together they topped off her fuel tanks and heaven only knows what else. The boat shop workers had left on board a complete set of new tools for the new engine. The last things considered were our survival clothing. The medic fitted them to our size and stenciled our respective names on each piece. An electrician had worked all night to be sure that the electrical systems were up to his satisfaction. Everything was well-organized and we were ready to be put on the water for another go at destroying mines and mine layers.

The ship's captain appeared and asked that the Sergeant and me to accompany him to the chart house. We all gathered around the chart table. The captain began by asking me if the boat were completed and ready to go on a mission, and then asked the Sergeant the same question. Both of us answered, "Yes sir, it is."

While pointing to a position on the chart he asked, "Sergeant are you familiar with this location?"

"Yes sir I am, that's the Yalu River basin."

"How about you coxswain?"

"I agree with the Sergeant, sir."

"Have both of you had experiences in this area?"

We both answered in the affirmative.

"Well, that river is the origin of the Sampans that are transporting the mines. It has been confirmed by aerial surveillance. Since both of you have been there, I would like for you both to plan an attack for us so they will know that we know where those mines are coming from. I will be moving at sixteen knots northwest along this course which I have indicated with a red marker until I hear from you."

"What is our present position, sir?"

"Oh, I forgot that. We are right here at this red circle."

"Thank you sir."

We had just begun when reveille sounded and we decided it would be best to have breakfast before we got too deeply into our assignment. So we began talking.

"What about this, Sergeant? We use our mother ship as a decoy."

"In what way?"

"Well, Russian submarines know where she is every minute and they also know that we are operating from her. So whatever mischief we create on land, they believe we will run back to her for safety, especially if she remains fairly close to our launch site. Instead of running away from them as they will expect us to do, we will run toward them, which they will not expect us to do. After we have made our attack, let's run south close to shore, with our North Korea flag flying, for two hundred fifty miles to the LST. It will be about a twenty-hour hull speed run.

"And there is one thing you can count on, that since our mother ship is beyond Russia's claimed defense line waiting for us, they will draw a straight line from where the attack took place to her position, and then bet it all on catching us somewhere along that line."

"Captain, how long have you been thinking about this plan?"

"Since we were told by the captain a few minutes ago to make one."

"I'll think on it a while but your plan seems to be the only way we can live through this attack, let's go have breakfast."

Chapter 37

We had a very good meal with our teammates and advised them to fine-tune everything on the boat because something big was in the making.

"Captain the more I think of your plan the more I like it, but I still have a problem!"

"What's that Sergeant?"

"Firepower! I see the mission as a great opportunity to destroy many Sampans that are loaded with mines but only if we have enough firepower."

"Sergeant, do you think we can borrow additional firepower from our mother ship?"

"What, for example?"

"One of their fifty-caliber AA (anti-aircraft) guns."

"Where would we put it?"

"My preference would be the bow because it is a smaller, more difficult target for our opponent to attack. Broadside fighting is a quick way to lose the fight, and win either a six-foot hole in the ground or a permanent residence in Davy Jones' locker.

"Sergeant I believe that with one good shooter, one good loader, and one fifty-caliber AA gun, we can do more damage to our enemy in two minutes than our present armament will do in a half-day fight. And with our fast boat, we could possibly be out of the target area before anyone realized what had happened. The rifle shooters could be used to protect our backs. Our lives will totally depend upon speed, surprise, and firepower."

"Let's have a look at one of those fifties, Captain."

We were less than twenty-five feet from one with no cover. "Wow! Just think what we can do with this weapon," I exclaimed.

"Captain, can this gun be mounted on our boat?"

"They put them on military Jeeps, which are much smaller than our boat. When gunner's mates are not close around, they usually cover their AA guns with canvas to protect them from salt air. There must be a gunner nearby because this one is uncovered!"

Sure enough there was. He had seen us looking at his gun and walked over to meet the assault boat leaders he had heard so much about. The Sergeant wasted little time learning everything he could about the gun before he asked the gunner, "Can a gun like this be installed on our boat?"

The answer was, "I don't see why not."

"How long would it take?"

"Well, let me think a moment about that." After a brief pause he answered, "My guess is three hours."

"Do you have a few minutes you can spare to have a look at where I would like to put a fifty-caliber on our boat?"

"Yes I believe so. Let me cover my gun and then we'll go have a look."

We climbed aboard our boat and the gunner asked, "Where do you want it put?"

"Well our plan is to attack from the bow with all of the firepower possible."

After looking about, the gunner said, "There is plenty of room and strength to put one or even two guns six feet aft (behind) of the Samson post."

Two! Now that really excited the Sergeant.

"I'm concerned about one, so how in the world could I ever come up with two!"

"Just ask is all you need to do. I have four in my back-up inventory and your assault team are now crew members on my ship so there is no problem."

"Well sailor, I'm asking! The target I'm after is two hundred miles north of us and we will have only one chance to destroy them. So the more firepower we have the better our chance of doing devastating harm to them and coming back in one piece. Our ship's captain has ordered us to develop a plan to attack the source of the floating mine problem and then report back to him. We have completed the plan, and will appreciate your meeting with our ship's captain and us to explain the fifty-caliber request."

"Sure I'll do that."

The Sergeant went to the bridge deck and returned to the chart house with the captain. Both of us, one and then the other, began explaining our plan. However, when we got to firepower, I yielded to the Sergeant, whom I believed to be a brilliant military leader. When he was questioned by our captain, he highly recommended the fifties, which concluded the explanation of our plan.

The captain responded that he agreed with us, "Except for one thing. The team needs a gunner's mate to make sure the fifties are operational when they are needed."

We all looked at our guest with hope that he would volunteer. He finally answered our pleading stares with, "Okay, I guess that's me!"

We all shook the gunner's hand and thanked him. He and the Sergeant left the chart house to begin work on installing the guns. The ship's captain reminded me that he would need a time schedule to go along with our plan. "Yes sir, I'll do that right away and get a copy to you!"

I realized that with no one to distract me, I could work faster on times, courses, distances, and an escape route to our plan. But it was only a brief thought, because without adding anything else there were still plenty of time-consuming things that I must yet complete.

The first thing I must do is assume when and where the largest number of Sampans loaded with mines will assemble before starting their trip to the south-flowing currents in the Sea of Japan. It is very probable they will group in sight of the white light which marks the entrance to the Yalu River, and because of their deadly cargo they will stay clear of each other by two hundred yards or more. They can not navigate after dark except on the river, where they must have an outgoing tide between six p.m. and six a.m., in order to reach the river delta light in time to depart with a gathering flotilla of forty Sampans, more or less. Each boat must leave from the "ice barrier" alone and travel single file at predetermined intervals for ten miles to the river's entrance light. There will be no patrol craft of any kind to govern their attention to safety standards, or to assist with their scrambled departure east at dawn, because their cargo is nothing less than mortal danger.

Chapter 38

I wrote our plan in two parts. The first part was given to our ship's captain so he would know exactly when, where and how to launch our boat. After we had been lowered onto the water and had moved safely clear, they would remain on station until one p.m., acting as if she were preparing to pick us up again at any time. We hoped that such an appearance would draw Russian search efforts toward our mother ship and away from our true escape route. And if we were very fortunate, that bluff would give us an approximately sixty-mile head start away from the search area.

The second part was given to the Sergeant so he would know exactly when and where my navigation would take us to a place where we would fight. In this scenario, if my assumptions proved to be correct, a large number of Sampans would be gathered about five miles southeast of a narrow channel which leads to the Yalu River's white entrance light. And ultimately it leads to a staging area for loading mines onto the fragile Sampans. It was just a guess but I believed the Sampans that had already made their ten-mile down-river run would remain together (as close as possible) near the Yalu entrance light in an approximately one square mile area, and hold back until dawn before any of them would put out to sea. We were reasonably sure where they would be waiting because we had been there before, and we were also reasonably sure when they would depart for the south flowing currents in the Sea of Japan. Another benefit we had was that since our last visit to the Yalu basin had caused such an international ruckus at the UN, our enemies would never expect us to do the same thing in the same place a second time.

Our assault team must arrive at my assumed location precisely at the first light of dawn if we expected to catch the gathered Sampans while they were still an easy target. And since it would be bitterly cold (below zero), the Sampan boatmen would certainly be heavily bundled up to keep from freezing to death. Therefore our approaching boat's engine noise would likely be unheard, or even ignored. With our added powerful guns (fifties) it was my opinion that we would complete our business with the enemy in two to three minutes, and then run south near the coast at full speed for two hundred and fifty miles until we reached the grounded LST in South Korea.

I took the captain's copy of our time schedule to the bridge and discussed it with him. He asked when and where I would like to be put on the water.

"Fifty miles southeast of the Yalu River delta at midnight tonight will suit me quite well sir."

"I'll confirm this with you within the hour coxswain."

"Thank you sir! Captain, permit me to say that if we have any hope of deceiving the submarines that are probably tracking us at this very moment, it's essential to the success of our mission that we launch our boat at ten knots. Because the Russians will think that it can't be done at that speed and especially at night. Therefore, our undetected launch will not cause an alert to their coastal defenses and/or to their ally, North Korea. Also the launch must be done with no lights burning on our mother ship except normal navigation lights. If they should be turned off, that would create suspicion. The submerged speed of their submarine is ten knots more or less but their radar is far-reaching and if they should detect any significant change in our ship's speed or course, that would be cause for alarm. But our assault boat has been radar sanitized, therefore when we are free from our mother ship their radar can't detect us.

"Thank you for your time, Captain! I'll be on the boat deck if you need me."

"Understood coxswain."

"Sergeant, I have completed our plan to attack the Sampans! However, I see that you are busy with the fifties."

He answered, "We'll talk later."

"Good with me but I must tell you that the launch will be at midnight tonight."

I got busy with the things I had to do before mess call. However, there was one thing I felt compelled to do first and that was go over my plan with Sam. He asked many good questions and for that I was thankful. Sam, Hawkeye and I had to be perfectly coordinated during the extremely dangerous launch or the plan would not be worth the paper it was written on, because with just one mistake we would all be fish bait.

"Sam, there is something that has been really bothering me."

"What is it Captain?"

"Do you remember that on our previous mission to the ice barrier, we saw an unusually large number of Sampans pulled up on the river bank?"

"Yes, that I clearly remember. They were on the Russian side of the river – neutral territory. Captain, are you thinking about going up the river again?"

"Maybe, because there are signs that those Sampans are being loaded with mines in Russia. Crossing the treacherous Yalu River to North Korea with such a sensitive cargo just doesn't make sense."

"Have you considered the tide?"

"Yes I have. We know the Sampans came to the delta on an outgoing tide, but we have no way to know how long the water has been flowing back to the sea, and I will not go on the river against an ebb tide."

"Have you thought about the entrance buoy's reaction to the flowing water from the river?"

"I have, but that can be useful only if the buoy is free-floating and anchored to the bottom, then I can predict the tide flow with about fifty-percent accuracy from the water's pressure on the buoy. However, if it is stationary it will be useless for my needs, particularly in the dim light of dawn."

Chapter 39

"What really bothers me though is that we will be less than ten miles from the target of a lifetime and we might have to leave it without firing a shot! It especially angers me when I know that the single purpose of those mines is to kill American sailors."

"Well, me too, Captain, but we must not forget the fact that we would be attacking a target within Russian territory."

"That my friend will be our leader's call! But if we could put a few shots into their mine storage depot, which I'll bet is by necessity of convenient loading near the water, the chain reaction of all those explosions would be so humongous that it could easily be blamed on a B-29 raid. Because no one in their right mind would ever believe that we had done it yet again."

"Sam, what kind of time estimate do you think we would have from the river entrance light to the ice barrier?"

"With the best tide condition, thirty minutes one way. With the worst tide condition, ninety minutes the other way. So my answer is, round trip two hours minus what ever mixture of favorable currents and tide time we would encounter. In other words don't set your clock using my advice." We both had a good laugh after that and retired to our individual work stations.

The team members, except Sam and I, were as busy as beavers with things the Sergeant had directed them to do, such as preparing for the increasingly cold climate we would experience while steadily moving north. I overheard Hawkeye explaining to the four new Rangers their responsibilities during the blacked-out, high-speed launch. He also told them how vitally important it would be for everyone, as expected, to give their best performance because nothing less could be accepted if we expected to grow old.

On my list of things to do, I rechecked the watertight integrity throughout the bilge area before supplies were loaded. That is always a time-consuming job but very necessary. One example of a small but critical thing to watch for during the bilge inspection is to make sure the drain plug is twisted tightly into its thru-hull fitting. After the seawater has been drained, the plug is more than often put aside until later. That failure, if not detected, would cause the boat to sink fast after she had been put on the water, especially with her heavy load. The supplies (cargo) would not be able to be moved quickly enough to correct that seemingly small but fatal mistake, because in a brief period of time it would be too late.

Another problem was that with no lights, and all hands very preoccupied with high-speed launch procedures, the boat would have surely been flooded beyond our automatic pumps' capacity to save her before anyone would have even have noticed. I never assigned the boat plug responsibility to anyone. I did that job myself. There are enough other things out there to kill us all without adding our own carelessness and forgetfulness to the list.

For us the war began the moment our boat was lowered onto the sometimes cruel sea and it ended the moment we were hoisted safely back onboard our mother ship.

"Mess call Captain!"

"Yeah, I heard and I'm ready."

Our team proceeded to the mess deck together, therefore our conversations were totally about making ready to launch the boat at midnight. While we were eating our meal, a messenger who had been sent from the bridge deck advised the Sergeant and me that the captain wanted to see us "right away." That was all we needed to hear! We rushed to the bridge where on our arrival we were informed by our angry captain that he had just received information from "flag" (fleet command) that a USN mine sweeper (U.S.S. *Pirate*) had been sunk by floating mines. However there was not yet any information about the number of casualties.

"But our orders are to put a stop to those damn mines!"

The Sergeant asked, "Sir, does that mean we have your permission to go hunting up the river?"

"Yes, Sergeant, you have my permission! Do you have the two fifties installed yet?"

"Yes sir we do!"

"Good! Don't forget to take plenty of ammunition with you because I want those bastards to know we have been there. Also remember to call on me for anything you need. I'll be here on the bridge or in my sea cabin until after you have made your dawn attack. The wireless code words for success will be Rose Bird. Coxswain, my navigator has plotted the northwest course that you will need to locate the Yalu entrance light and the exact distance to it from the time you pull the sea painter pin. He will give both of them to you just before you go over the side."

"Thank you Captain."

He then turned to the Sergeant and ordered, "Sergeant, from the instant your assault team is free from the sea painter, it is completely your decision about anything you do until you are back on it again. Do we understand each other?"

"Aye, aye sir. Captain, I'm certain that we do."

"That's all for now, gentlemen!"

We hurried back to the boat and continued preparing for the dangerous launch. On our way back the Sergeant advised me not to be surprised if he called upon me to take the team upriver to the ice barrier again, because that's where the prize targets would be found! After I had shared that news with Sam he looked at me, with his I-told-you-so smile, without saying a word. Sam and I hand-signaled Hawkeye and he came to the cockpit. The three of us decided that it was time to begin boat launch practice drills, because if we didn't get this one perfect, the rest would be a waste of time! I called for a meeting of our team at the steering cockpit for launch practice. After we were together, I began by warning everyone that there must be absolutely no talking during the launch exercise.

Chapter 40

"The reason for that order is to prevent communication between the captain (me) and the deck launch officer from being jeopardized by unnecessary chatter which could unintentionally cause a fatal mistake. Launching a MWB at ten knots in total darkness is a very frightening experience but if everyone performs their duty exactly as they have been trained, we can safely make the launch in less than two minutes and then be on our way. At this time I must make a correction to what I have previously said in error, the Sergeant and I will be the only people who will speak during the launch. He well knows the responsibility at each man's launch station and he will make sure that all commands from me are clearly understood by each of you. It's vitally important for each man to clearly know that once the launch exercise has begun, there can be no turning back and if a mistake is made, there will be no second chance. We must get it right the first time or we will without a doubt be cut into fish bait by the ship's propellers. The sea is a little more restless than the time before when we had plenty of daylight. However, the Sergeant and I remain confident that our team is ready to prove they can also do it with no light. Due to deteriorating weather conditions, we believe the Russian and the North Korean border patrol people will temporarily drop their guard, because they more than likely believe that only a very foolish sailor would even think about lowering a boat, at any speed, onto such a stormy sea in below-zero weather. The Sergeant has taught me many very important things about being a Ranger, not the least of which has been to go where the enemy least expects you to be, and attack him when he is least prepared for you to do it! As your boat captain, that is exactly what I have been ordered to do. He also taught me that to be undetected (surprise) is at least half of any battle.

"Now you know that my choice – to use a very dangerous high-speed launch at midnight with darkened ship conditions – is a sure way to catch our enemy sleeping. The Sergeant and I know where they are, and we know when to best catch them off guard – at dawn."

I ordered Hawkeye to put the team through drill after drill while I without any mercy supervised them. They seemed to have difficulty understanding why I would not release the sea painter until their body weight had been shifted aft. I gathered them around me and began to explain thusly.

"The last and most dangerous assignment to be carried out during the high-speed launch is to pull the painter's release toggle. After having done that, if by chance the boat's proper balance favors aft (stern), which means the bow will run high on the water instead of plowing low beneath a wave, that captain has been blessed with pure luck. But your captain does not depend upon luck. I will not on my boat order the painter's toggle pin to be pulled until she is in every way ready for whatever sea condition that might challenge her. For example: Your combined body weight, seven hundred pounds or more, moved quickly to the engine area just before the painter is released, will greatly improve my chances of successfully executing a not-in-the-books exercise. Please remember that my success as your captain will be your success in battle! But first I must get you there! Our Sergeant will tell you when and how to move aft to the engine spaces."

"Are there any questions? Since there are none, let's continue the launch drill."

The Sergeant signaled for me to join him on the main deck where he was standing with the chief boatswain, who would be our launch officer. When I arrived, the chief expressed to me his growing concern about the dropping barometric pressure and the increasingly cold wind, which was sure to get even colder after dark.

The Sergeant asked me, "What about that, Captain?"

"Well it's hard to predict the weather but I do know that our enemy is in a place where we can destroy him at first light and that is what I want to do if at all possible! However, I'm not a suicide coxswain! I truly believe that our assault team is razor sharp, prepared and ready for our boat to be put onto the restless sea which I have been observing for the past several hours. I would not go over the side under these conditions with just any team, but this one is truly special and with their dedicated help I will, at dawn, coxswain them eyeball to eyeball with their enemy who has killed many of our brother sailors with Russian floating mines."

"Coxswain, do I inform the ship's captain that we have a go?" the chief asked.

"Please tell him that for now we have a go. I will continue to monitor the sea conditions and give him a final decision at eleven-thirty."

"I will do that right away, coxswain, but permit me to say on behalf of your mother ship's crew, this team is the best thing that has ever happened to this ex-mail-carrier ship and we don't want to lose any of you."

Both the Sergeant and I thanked him kindly for that statement. I reminded the Sergeant that the sun was sinking low and we must complete our loading before dark.

"Aye, Aye Captain, I've already completed the organizing, and we'll have it all on board by sundown."

"That relieves me very much, do you plan to deploy the shelter for a few hours before the launch?"

"Yes I do, and I have gotten our deployment time, up or down, reduced to five minutes."

"That's wonderful, I hope Sam's engine heaters are up to his expectations."

"So do I, and I also hope the low-voltage night lights can be used under the shelter without detection. But I will make sure about that Captain. However, most of the new additions have worked out quite well for us!"

Chapter 41

"Does that statement include the fifties?"

The Sergeant laughed about that and answered, "I'll let you know about those babies in about fifteen hours."

"I hope that what you tell me at that time is, 'There are no more Sampan targets left to shoot.'"

"So do I, Captain, so do I."

Evening mess call was announced over the PA and all workers who had been helping to prepare our boat for the midnight launch assembled on the boat deck. I thought this a bit unusual. We all began the short walk to the mess deck. After I had been subtly pushed ahead through the entrance hatch, the whole place went bananas. Every off-duty crew member and our ship's captain, who is never off duty, were there to wish us well. A special meal had been prepared and with everyone holding hands the captain offered up a beautiful blessing which included a prayer for our safe return. The gathering lasted about an hour after which our team thanked everyone and headed back to top off (complete) our boat for the midnight launch. When we had returned I noticed that the chief (deck launch officer) and his launch crew were checking everything twice, so I took that opportunity to inform him about something that had been troubling me for some time.

"Chief, the davits on this small vessel are rigged in such a manner that it could be very dangerous for us during the high-speed launch."

"How is that coxswain?"

"Two of the eight monkey lines are much too close to their respective lowering/hoisting blocks (falls) and are not properly secured. If one of them were to come loose, it would fall directly into the block below it and surely disable it. We could not survive that mistake."

After carefully observing, the chief responded, "You are absolutely correct, coxswain, maybe you have a suggestion for fixing the problem so it will not be necessary to remove the spreader cable from the davits?"

"Yes Chief, I do. As I have said, I've been studying the situation for a considerable number of hours and have come to a simple solution."

"Let me hear it!"

"Remove the two monkey lines that are closest to the blocks. That will leave six, which are enough to meet Navy requirements for a twenty-eight foot whale boat."

"I'll put my men on that problem right now! Thanks coxswain."

"There is something else about the blocks, Chief."

"Tell me about it."

"All right, as you already know, immediately after the heavy fore and aft blocks have been released from the boat, they become very dangerous free-swinging hazards which cannot be seen during the dark hours. However, if a line (rope) could be attached to each of those hazards and tended by two men (one tending each block) on the boat deck, that problem would be history in no time at all."

"Very good idea coxswain, consider it done. If I let anything go wrong with this mission, the captain will probably have me keel-hauled (punishment for a careless sailor who endangers the lives of his shipmates) for sure and if I should survive that event I will be a seaman apprentice until I retire."

"Well Chief, he badly wants those Sampans so let's see what we can do for him!"

"Right coxswain, let's make him happy!"

The wind from the storm had increased to forty-six mph and rising, and the seas (waves) were estimated to be five- to seven-foot. That weather update was reported to me from the bridge, along with a recommendation from the chief launch officer that I abort the mission. My answer was negative to abort, unless ordered.

The chief handed to me, in very dim light, a water-tight container and shouted, "Your orders, coxswain!" He also shouted that we had reached the position and speed which I had requested to make the launch. Shouting through headphones is not always comfortable but I had understood everything that was said.

"Very well Chief, report to the captain that I wish to go over the side in fifteen minutes, because waiting in this storm will only increase the danger. Please do not use the green light, instead use headphones only between you and me until I'm free from the painter, and then use VHF channel 13. Take care to keep my bow three feet higher than my stern, and make sure that you point my bow into the wind."

When the time came to go, I began readiness reports. The Sergeant was first to report.

"Ready in the boat Captain!"

"Sam?" (It wasn't necessary to tell him anything.)

"Ready Captain!"

"Ready Hawkeye?"

"Ready Captain."

"Chief, put my boat on the water slowly and please don't let us hang on the davits with no water under us!"

"Aye, aye coxswain, I'll make sure that does not happen!"

When I felt the lowering of my boat, I knew the exercise had begun and ordered the Sergeant to stand by to release the aft fall (block). When I was sure the time was right, I spoke a little beyond normal into the head phone, "EXECUTE."

The Sergeant, using his headphone, reported back to me, "Aft fall clear." That pleased me very much because those two sailors on the boat deck had tamed one of two extremely dangerous mistakes that had been made while rigging the davits!

Now for number two, I thought. Sam was the key to our staying alive with this one. He must not let our boat fall back too quickly on the painter because if that happened we were goners. The sudden and severe jerk on the Samson post would surely cause it to give way and open our bow to the sea.

"Stand by to release the forward fall!"

God please send me the best wave you have Right Now because I can't wait any longer.

Suddenly my boat's bow was lifted on yet another very promising wave.

"EXECUTE!" Perfect again, I thought! Now for the meanest, most unpredictable event of the entire exercise, the sea painter.

Chapter 42

"Sam, with this strong headwind will you be able to keep up with our mother ship after we are off the painter?"

"Yes Captain, in fact I'm doing that right now with power to spare."

"How does the tiller feel?"

"As long as she stays equal to or just a touch faster than the painter tow, I can control her well enough, but when she falls back on the painter that puts heavy down pressure on the bow and controlling her then becomes extremely difficult."

"I'll take care of that for you Captain."

Silently, I resolved to get off the painter as soon as possible. First I'd better test headphone clarity with Hawkeye to make sure that he could understand me when I ordered him to pull the toggle which releases the painter. Because pulling it at the right time is a unique feeling that every successful coxswain must learn, and he must instantly relay that feeling to his bowman. That's why I put Hawkeye on standby until that special moment challenged me. I heard Sam and the Sergeant softly whispering into their headphones – "Give the order Captain, do it now!"

"Not yet mates, not quite yet," I whispered back, and then what I had been waiting for came to me like a shot. "EXECUTE!" I shouted.

Several members of the team shouted, seconds after I had, "PULL IT Hawk!" That delay alone should convince any reader that our boat had been lowered onto a very rough sea and controlling her was no easy task. Sam quickly matched our speed with the mother ship and struggled to maintained that speed until we were a safe distance from the steel hull. Once there, yet another struggle was waiting for him and for me. That struggle was to match the sea's storm profile with our boat's motoring capabilities. I gently applied about five degrees of port rudder and slowly moved her away from the ever-present fear of our mother ship's propellers. There are many ways to die but being chopped to pieces by a ship's spinning propeller is on the bottom of my list of choices.

The deck launch crew had obviously hauled in the slack painter and announced to our shipmates that we had made it. Even though they could not see us, they were deeply concerned about our safety. Loud cheering erupted from the darkness, but this time the Sergeant had communicated with our mother ship and reported our circumstances to her captain. He was very pleased to learn that we were all safe after having completed the dangerous launch. Our team was so well programmed that they reminded me of bees in a hive.

Even before Sam and I had coordinated our efforts to control our boat, the team had put up the spray curtains and the shelter, and in darkness no less! Then they began preparing for a very important battle which was only hours away. The Sergeant approached me out of the darkness and said, "Captain, your chart table is ready and the chart light is on. As soon as you and Sam have stabilized our boat enough so I can handle the tiller, our mother ship has navigation information for you."

"Thank you Sergeant, I'll run before the storm until I find a good heading for you to relieve me."

Sam and I eventually agreed on a course and speed to make a safe tiller transfer. When the Sergeant touched the tiller he asked, with no little surprise, "Where is the tiller insulation?"

I answered, "Under my feet. I needed to feel my tiller during the launch so I removed it."

"Well I'll put it back because I don't intend to take you into battle nearly frozen to death like I did before. However, that reminds me of something."

"What?"

"Your Navy doctor instructed me to look after you or I would answer to her. That means that I must grease your arm in an hour or so."

"Thanks friend, I'm ready for you to relieve me now, sing out if you need me. Sergeant, what would we do without each other?"

He chuckled and answered, "Probably not very well, Captain."

We were running before a bad storm, so when I unzipped the canvas to enter the cabin, very cold arctic air temporarily chilled the occupants. I was surprised that Sam's engine heaters were as effective as he had hoped they would be. The first thing I did before working out a course and speed to the Yalu's entrance light was to open the canister which held the Sergeant's and my orders.

Our orders told us to do anything necessary to destroy mine-carrying Sampans, without any regard for borders or civilians. However, we must not knowingly attack civilian targets. Again we had become ghost soldiers.

Now how in the world am I supposed to interpret a yes/no in the same sentence?

"Sergeant, call Hawkeye to relieve you, tell him to steer NW, then come to the chart table for your operational orders."

"What's up, Captain?"

Before we began our talk, I turned off my headphone. The Sergeant sensed something was up and also turned his off. "I have read my orders, now it's your turn to read yours!"

After he had read them, he said, "These are not orders they are a death sentence. Do they really believe we should approach a Sampan in hostile waters and inquire if there are any innocent civilians on board! The only way to win out here is to kill them before they kill us, and that is exactly the way I intend to fight our war. The brass (high ranking officers) need us to do a job for them which they cannot do for themselves because they must obey UN rules of engagement. But out here those rules, if obeyed, can only get us killed because they tend to favor our enemy. It seems to me that we are even more expendable now than before. And just as before, in case we are killed or captured, there will be no records to verify that any of us have ever existed."

Chapter 43

"Consequently, we will have no rights as corpses, and even less as prisoners of war. But anyway, that is probably an inappropriate joke because here in no man's land, prisoners' rights are totally unheard of – because the communists take no prisoners. After fighting long and hard using civilized guidelines and the world's finest military hardware, the brass still can't get the job done without using people like us to protect their political rear ends."

"And that, Sergeant, is very strong proof that your kind of fighting is the only way to win and stay alive up here. I seriously doubt if a single one of these primitive, starving peasants understands what a prisoner of war is!

"There is also considerable doubt in my mind that even one of our commanders understands how to fight these people. For example, give a starving man a cup of rice and he will do anything, such as transport mines in a flimsy Sampan which could easily cost him his life."

"Right you are, Captain. It seems that we are pawns – but with one exception."

"What's that, Sergeant?"

"I will not leave the fight until we are completely out of ammunition or until there is nothing left to destroy. We were sent up here to destroy Sampans and, by the saints, that's what we are going to do. Are you with me, Captain?"

"You bet I am, Sergeant! But when the ammo runs out, I intend to take you on a boat ride you will never forget!"

"You have already done that!"

"Okay, you will get another never-to-be-forgotten ride! While we are inside an almost warm cabin, it seems to be a good time to keep the promise you made to my Navy doctor."

"Yes, I'm glad you reminded me of that obligation."

After my arm had been properly cared for, the Sergeant called for condition "RED" which means make all preparations for battle. From that moment on I concentrated on my primary responsibility, which was to have my boat and her Navy crew ready to fight when ordered to do it. Hawkeye would always keep a sharp lookout for targets and for danger. Sam would top off the fuel tanks and make other things ready that only he knew how to do. My reputation as an assault coxswain was depending upon the fact that I had assured everyone that I knew where to find the Sampans at the break of dawn, and the Sergeant concentrated on making sure we were ready to fight, using his battle plan and the weapons he had chosen.

"How much time is on the engine clock, Sam?" I asked.

"Three hours and forty-five minutes since Hawkeye pulled the toggle that freed us from the painter."

"Alert me at four hours and thirty minutes engine time."

"Aye, aye Captain."

"Remember, I'm using only latitude for my navigation but I prefer dead reckoning for speed! We must not arrive at our objective before dawn, but since we have been running before the wind, it may become necessary for you to slow us down with our nice new sea anchor."

"Will do, Captain.'

"Hawkeye!" (on the headphone)

"Yes Captain."

"It is my intention to put the Yalu River's white entrance buoy light dead on your bow in about thirty minutes. Do you remember the height of that buoy?"

"Ten feet."

"That's correct sailor. Use it when you calculate our distance."

"Will do Captain!"

"Sergeant?"

"What is it Captain?"

"We are on a NW course and I expect to sight the Yalu's white entrance buoy light in about thirty minutes."

"How far after that?"

"Since the weather is moderating, my guess is forty-five minutes."

"I'll have our team ready to shoot as soon as Hawkeye sights targets. Agreed?"

"Agreed. Please have everyone use ear protection because this one will be dangerously loud. Also be mindful that the closer we get to a target the higher the shrapnel risk."

"Will do Captain!"

"Do you have your cyanide pill?"

"Yes I do. I removed it from your survival suit while I was greasing your arm and could not remember replacing it."

"You didn't, I did! When will you take down the shelter?"

"Right after I see the buoy light so that I can place shooters with BARs (Browning Automatic Rifle) at stern, port and starboard. Our gunner's mate and I, assisted by two loaders, will shoot the fifties."

"Barely visible white light dead on our bow, Captain."

"Keep your attention right there Hawkeye because that is where the fight will begin in a little more than thirty minutes. Sam, did you hear Hawkeye's report?"

"Yes I heard it."

"Can you two get us to that light by dawn?"

"Yes, we can do that, Captain."

After hearing that, the Sergeant began making ready for a battle that was only minutes away. "Take down the shelter, men! Man your battle stations, check your weapons and your ammunition and be prepared to fight like United States Army Rangers! Sorry – and four United States Navy men."

"Sergeant don't forget our battle prayer!"

"Thanks for reminding me, Captain."

He really did himself proud with that prayer. We were ready!

Hawkeye reported, "Sampans sighted on the port bow, Captain."

"How many?"

"Three!"

"It's not quite light enough yet, Sergeant, let's hold back about fifteen minutes until we can see how many Sampans we are dealing with."

"I agree with that, Captain. When you can see ten or more I recommend that we attack!

"I can see that many right now so let's go get 'em. Full speed ahead Sam."

"Aye, aye Captain."

"What is the kill range, Captain?"

"Five hundred yards."

"That will be your call, Sam! I'll steer clear but it is going to take both of us in a nest of targets!"

Suddenly every shooter on our boat had begun firing his weapon at the Sampans, and mine explosions became too rapid to count. Every mine transport we destroyed must have been carrying two or three mines. As I steered around the entrance light's wide expanse, there was no let-up at all in the explosions.

Chapter 44

The first time around, we destroyed, without a doubt, at least forty or more mine-laden Sampans. My plan for the second run was to make a larger circle while seeking out and destroying every survivor of the first attack. Then if circumstances permitted, I would head up the river and begin looking for the target of all targets – the mine storage and loading area.

On the second time around we located and blew up twenty-seven mine-carrying boats. Of that I'm sure because that time I could count the Sampans we blew up. The black smoke from all of those mine explosions is beyond my ability to describe but I will take a shot at it. Plain and simple it collectively reminded me of a volcano! But my great reward was knowing that those destroyed mines would never harm my shipmates.

"Give me all the power you have, Sam, I'm going up the river."

"Did you check the light buoy for tide conditions?"

"Yes I did. We've caught a mid-tide that's strongly rushing in." "Whoopee, I'll have you at the ice barrier in thirty minutes."

"Did you hear that Sergeant?"

"Yes I heard. Hawk, when you locate the mine loading dock, I'll go ashore and complete what I want to do in five minutes. Listen up people! I want everyone to hear this, the Rangers will be using timed explosives that are set for fifteen minutes each. If we locate their armed-and-ready mine storage area, the resulting explosion could easily blow us out of the water a mile downriver. Even fifteen minutes will be cutting it close, but adding the up-river time to our original plan gives us another big problem – too much daylight exposure."

The shooters remained at their battle stations as we sped up-river but they were instructed not to shoot unless cleared to do so by our leader.

Hawkeye slowly began to report, "I think I see what we are looking for."

"Rangers, stand by to go ashore!" our leader ordered. "Keep reporting, Hawk," he added.

"There are many Sampans on the Russian side of the river that are pulled up onto the river bank. Also there's a loading dock and an apparatus waiting to load mines onto Sampans."

"Don't touch any mines on the dock, Rangers, they may be armed!" the Sergeant warned.

"We've hit the jackpot Captain!"

"Great work, Hawkeye," I responded. "Now stand by the bow while I do a fast one-eighty and make my landing so we will be headed downriver, prepared to get out of here fast when the Rangers return."

Sam kept the river current neutral for us with the engine while Hawkeye and I held our boat tightly against the dock while waiting for the Rangers to safely disembark. All was quiet for about a minute and then all hell broke loose. This attack seemed to be much louder than our first raid on the North Korean side of the river. Grenades and gunfire continued for what seemed to be forever, but in fact it was only about five minutes. The Sergeant, with only five men counting himself, burned every building at that small work camp, and placed twenty timed dynamite packs on three large, separate stockpiles of Russian floating mines. I could hear them running toward the boat so I shouted to Hawkeye, "They're coming back, pull the boat in real tight against the dock."

The Sergeant yelled, while his feet were still in the air over the boat, "All Rangers are onboard, Captain!"

Sam and I quickly got underway before anyone had even sat down.

"Six minutes and twenty-one seconds Captain and not a shot was fired at us!"

"How much time until the big one?"

"Ten minutes, give a few seconds either way."

"All hands, this is your captain speaking, batten down all loose cargo and keep your heads as low in the boat as you possibly can!

"That includes you and me, Sam. I'm going to temporarily turn upriver and hold our boat at the lowest possible speed to maintain control. Then I plan to point her bow, the area of least resistance, into the expected explosion's wind blast."

Those were the longest ten minutes I had ever experienced but when they had finally ticked by, I heard the most indescribable collective explosion of the entire Korean War, and we were less than one mile from ground zero and near high tide. My eyewitness account of these bizarre events which I'm about to write is something that people just won't believe! Anyway here goes.

At near ground zero the sub-zero temperature briefly became shirt-sleeve warm and a three to four-foot wave which had been caused by the blast was formed on the river and came charging toward our bow as I had predicted. When it hit our boat, the impact gave us all a big fright, especially knowing that if our boat had capsized in that frigid water we were as good as dead. I quickly regained control of my knocked about boat and chased the fast-moving wave as long as possible downriver to reduce tide resistance. Some of us may never properly hear again, but who cares, we're still alive, I thought. Thank God I had put my boat in a survival mode before the explosion!

During my six years at sea, I had two extraordinary experiences that I have never been successful at getting anyone to believe what I was telling them. So

I stopped trying. The above event was one of them, and a southwest Pacific typhoon, in which I was trapped for three days and four nights, was another.

"Give me all of the power you have Sam because we will be bucking a four-foot tide all the way to the entrance light!"

"Aye, aye Captain!"

"Sergeant, it's conference time!"

"Be right there Captain!"

"Before I begin, Sergeant, I must compliment you on a job well done."

Chapter 45

"With your amazing military leadership, both of us got what we so desperately wanted."

"Indeed we did Captain. Have you a plan for our escape?"

"Yes, but it is filled with doubts. First, there are no helicopters in sight, which could be easily seen if they were searching for us, and that is a big puzzle for me because we are on the river in clear daylight. Next, consider the tide facts, we are nearly a one-hour boat ride from here to the river's entrance buoy, which is our only gateway to the open sea. Since they know that we are bottled up and can't go any other place, they must be waiting for us at the entrance buoy, where they can easily blockade the river and assemble every weapon they have just for the pleasure of shooting us. But how could they have gotten where they are without having seen our boat enter the river? The best way they can destroy us now is with helicopters but there are none in sight. There must be another Sampan anchorage! God help us, we must have missed some! They're going to blockade us with floating mines. That's our situation Sergeant, I recommend that you prepare for a tough fight at the river's delta. How much ammunition do you have left?"

"I still have enough to put up a pretty good fight but none to waste."

"Does that mean we have fifty-caliber ammunition left?"

"Yes it does Captain. It also means we have only a little bit of dynamite, and even less grenades and BAR ammunition."

"That's not good to hear! But whatever we encounter at the entrance buoy, make sure that every shot counts."

"When the time comes, my plan will be to take out enough Sampans to make a hole big enough for us to get through. And then go after the ones we missed."

"One other thing Sergeant, I do not believe there will be any armed personnel even close to the explosive blockade material, because that would be nothing less than suicide. When we get to the delta, I will wait until the tide is completely slack and then steer as close to the blockade material as safety will permit. At that time, Sergeant, it's our Rangers' expertise that will save the day. Slack tide will give you thirty minutes and no more!"

"Aye, aye Captain, get me there as soon as you can so I can observe the blockade material."

"Did you hear the Sergeant, Sam?"

"Aye, aye Captain."

"Well let's go do it!"

When we arrived at the blockade, the tide was close to slack, perhaps thirty minutes remaining. The Rangers had already formulated a plan of action. Even though it was snowing hard, the Sergeant called for the deck to be cleared, which meant lower the shelter. Next he stationed four men with our longest boathooks to fend off any material that might float in our direction.

"Captain there are ten Sampans blocking the channel, and you can bet your last dollar they are loaded with mines. The problem is I don't know where those mines are loaded on those boats and I don't have enough ammunition to spray fire and find out. But if we can explode one they will all go. And that creates an even bigger problem because if we explode ten or twelve mines at one time, we will need to be at least a thousand yards away to be safe. If we only had enough fifty-caliber ammo we could spray fire and take out those blockade boats in no time. But whatever we do it must be done in a hurry!"

"Sam, put us in the neighborhood of one thousand yards away and I'll point our bow at the blockade! Sergeant, you and Guns please meet with me in the cockpit. I know what you were about to do Sergeant but I couldn't let you go aboard those Sampans with your timed dynamite because you are too important to our team! Anyway, I have an idea that I think will work."

"Let's hear it, Captain."

"How many rounds (bullets) of fifty-caliber ammo do we have, Guns?" (nickname for our Navy gunner's mate)

"Exactly fifty-seven, Captain."

"Can you fire as little as ten rounds per burst?"

"An experienced gunner could do it but I can't promise that I can!"

"Okay, here's the plan. On the Sergeant's command, we will shoot the first burst at the blockade from a distance of one thousand yards. If we fail to explode the target the first time, we will move one hundred yards closer to the blockade and repeat that strategy, until we have no more bullets.

"Always shoot at the same Sampan, we don't have enough ammo to hunt. Now what I have said is an idea, not a command. I have two expert gunners on my boat who can take the ball and run with it. Gentlemen, make a big enough hole for me and I'll run our boat through it so quick it will make you dizzy. But whatever the case, I will continue on a southeast course until someone or something stops us."

"May the Lord's will be done, Captain."

"I agree with you, amen!"

The tide was beginning to flow back to the sea, therefore it was becoming more and more difficult to maintain my thousand-yard standoff from the blockade. But Sam and I remained vigilant while waiting for the first shots. The Sergeant signaled all heads down and fired the first burst. No explosion! Sam and I closed on the target by one hundred yards. The Sergeant held up two fingers, signifying he was prepared to fire the second burst. I rocked my chin giving my approval. Again a signal for heads down was given, after which he fired again. Still no explosion. He approached me at the cockpit and said, "Captain, maybe we should try a different boat!"

"I don't think so, Sergeant. Because I think I have just figured out the answer to our problem."

"Well please tell me what the answer is!"

"Sergeant, I have from the beginning, wondered how two half-starved Chinamen could unload a four-hundred-pound armed mine from a Sampan into the water. Now I know the answer. They float them out from beneath the Sampan."

Chapter 46

"There is no time now to tell you how it's done, just believe me and aim your next burst below the Sampan's waterline and you will hit the mines!"

"All of our fifty-caliber bullets?"

"Yes, all of them!"

"The same boat?"

"Yes, the same boat!"

"Do you intend to move our boat again before we shoot?"

"No, that will not be necessary."

On the shooter's next try there was complete success, a huge explosion occurred that almost blew our boat out of the water and opened a hole in the blockade big enough to run our mother ship through! To our amazement, after we had sped past what was left of the entrance buoy, we saw no one anywhere who was prepared to oppose our escape to the sea! Where are the Russians? I can't believe what I'm seeing, or not seeing, I thought. What we had expected from Vladivostok was aircraft, but I can't imagine why that did not appear. I continued motoring full-throttle away from the river's destroyed blockade. Not a shot was fired by anyone! Notwithstanding the fact that we did not have any ammo to fire!

"What are you going to do now Captain?"

"I'll tell you as soon as I catch my breath and can talk again, Sergeant!"

For several precious minutes the entire team looked all around us for enemy surface patrol craft while I varied my course a little to improve their field of search. When no enemy had been seen, I decided to run SE as fast as I could, expecting at any time to be blown out of the water by Russian helicopters.

"It has begun to snow real hard, Captain, should I put up the shelter?"

"Yes, the snow will help camouflage our boat."

I could not help being pessimistic, because I continuously expected an attack by North Korea or Russia at any moment, and we were out of ammunition.

"Sam, make a time note from the engine clock estimating the time we passed the destroyed river entrance buoy."

"Aye, aye Captain!"

"And while you are at it, I need fuel for three hundred and fifty miles at full throttle."

"I'll need to get back with you on that one, Captain!"

"Thank you, Sam. Conference time, Sergeant."

"Be right there, Captain. What's up?"

"My current position report is ready."

"Go ahead when you are ready!"

"Don't forget to send "Rose Bird"."

"Thanks for reminding me."

"I've been steering a southeast course since shortly after leaving what is left of the Yalu River entrance buoy. That course will intersect the thirty-eighth parallel which is estimated to be three hundred miles southeast of my present position. Then I will turn west, motor fifty miles, and rendezvous at the grounded WWII LST with people from our mother ship. Our estimated time of arrival, based upon twelve knots headway, is forty hours from my present location. Our fuel is critical, therefore I must be refueled by noon tomorrow, Sergeant!

"That's it for now except the weather, and you can report that. Sam will give me a current fuel report as soon as he can, but I don't expect it to be good news. For now everything is all quiet and secure because this snow storm is protecting us from air attacks. However, after dark I will feel much safer.

"I would like for Hawkeye to alternate tiller watch with me, two hours on and two hours off. Sam always sleeps in his seat, as you already know, and you can make whatever arrangements that you feel necessary for the others. I will say that the most dangerous time for us is between right now and dark."

"Captain?"

"Yes, Sam."

"We can make it to our rendezvous with sixty percent throttle and no deviations."

"Did you hear that Sergeant?"

"Yes! But it's your call, Captain."

"Let's go with one hundred percent during the night hours, Sam. I don't want to be unable to chase a target if an opportunity comes along. So let's pray that we are refueled by noon tomorrow."

"Aye, aye Captain."

"Now Sergeant, you can report that we will continue motoring at twelve knots because it is not safe in these waters to do less."

"By the way Captain, our gunner's mate ("Guns") brought along some coffee, I'll fix up a batch if the engine is hot enough."

Sam spoke up and said, "Don't you worry about that. It's hot enough!" Guns had also brought along a pot and twelve tin cups. Heavy snow and approaching darkness would protect us from those who would do us harm and our warm cabin will protect us from the arctic cold. So for the time being, I relaxed a little bit, if that is possible for a spy behind enemy lines in the world's cruelest war zone.

Snow began to come down in an abundance of beautiful large flakes, so much so that the shelter began to sag. That problem was soon taken care of by the Rangers. But they did not remove the snow from the shelter's top, instead they raised the apex of the roof so that it would slide off on its own. I had learned the benefit of insulating with snow early in my career with the Rangers.

The anticipated coffee was soon brought to me and at that particular time in my life, nothing could have been better! It had become totally dark so I ordered all navigation lights to be turned on and they made quite pretty reflections off the stormy snow. But I do not believe that another navigator could have seen them from fifty feet away. We were locked into a storm with forty-knot winds blowing out of the northeast. That was to our advantage, because those strong winds blowing from my stern quarter gave me a tremendous sailing opportunity. In fact they were so good that I had to order Hawkeye to deploy the sea anchor to slow us down! Sam didn't need to be told to reduce engine power.

I had never seen such large snowflakes in my life.

Chapter 47

As the flakes fell among our red, green and white navigation lights, I was reminded of Christmas. I noticed that everyone had come outside of the shelter to see a very reminiscent scene. Nine far-from-home warriors on a stormy night at sea had, by the miracle of Jesus Christ's birth, become little boys again and each was remembering celebrations of the past honoring His birthday.

We all remained quiet for a while listening for the bells – not always in vain! But I'm sure there were nine silent prayers being offered up. I'm also sure that all men, at some time in their life, yearn to be a little boy again because those were uncomplicated, happy days with families and friends that can never be forgotten.

Finally, everyone except those on watch returned to the warm shelter and soon thereafter to their precious dreams of home and hearth. The Sergeant had assumed the responsibility for watch rotation and it goes without saying that it was fair, well-organized, and timely. Our leader spent his spare time communicating with HQ on our wireless because we were too far out of range to use our radio (VHF).

During the night the storm seemed to have increased in strength and it was becoming difficult for me to manage the tiller. Therefore, I ordered Hawkeye to slowly lengthen the sea anchor drag line by an additional fifty feet, which would reduce our speed and stabilize the tiller. After that it could be fine-tuned by pulling the drag line in or letting it out a few more feet, to learn by feel the optimum combination of boat, wind and sea. To say that a sea anchor is a life saver, and should be part of every boat's safety inventory, is as close to the gospel truth as anyone could ever be.

I was teaching Hawkeye the techniques of sailing before the wind, using only the canvas shelter for a sail, when suddenly we collided with a whale, I guessed, which abruptly woke everyone, threw Sam out of his seat, and Hawkeye and me hard onto the cockpit deck. Immediately I ordered our leader to inspect the entire inside hull for watertight integrity. It was best to keep her sailing on course, I thought, because if there were serious damage, it would show up quicker with the hull under pressure. Water was discovered coming into the boat around the Samson post and the Rangers jumped right on it, giving full attention to the damage and the accumulating water. Sam made ready to use his large bilge pump and for the moment I maintained my course and speed.

It was four a.m. and I put Hawkeye back on the tiller while I determined our position for transmitting to rescuers, even though the Sergeant had assured me that he had the collision situation well under control. But things like that seem to grow as time passes, so I began all preparations for our rescue. "Sam, give me the engine time from the destroyed river blockade to here, and also our average speed."

"Sixteen hours at twelve knots, Captain!"

"Thank you Sam. Sergeant, report to HQ that we have collided with an unidentified object but we think it was a whale. We are motoring southeast at twelve knots, but I cannot at this time predict how long we will be able to sustain that speed, because we are currently taking on water at the bow and our hull is taking a real pounding from a snow storm! So as captain of this vessel, I am officially declaring, right now, a mayday!

"Also remind HQ that we have important, confidential documents on board. Our current position is approximately two-hundred miles southwest of the Yalu River entrance buoy and fifty miles north of the fortieth parallel. Send it, Sergeant!"

"Aye, aye Captain!"

"Hawkeye, have I taught you how to rig a collision mat?"

"Yes you have, Captain."

"Please go forward and stop the seawater from coming into our boat."

"Will do, Captain!"

I'll try to explain for the reader the purpose of a collision mat. Imagine if you will, an ordinary doormat with long ropes tied to each corner. Place the mat over the damaged area outside of the hull and then pull the corner ropes tight. That will stop most of the incoming water and slow further weakening of the damaged area. However, material to do the job must be carefully chosen.

"Captain, I have good news. Our mother ship has been waiting for us at about the same place where she had put us on the water. Her current position is about one hundred miles northwest of us and she hopes to rendezvous with us in about five hours. However, her captain advises that if both vessels steer on collision courses, much better time for the rescue can be accomplished. Please advise about your acceptance or rejection of collision courses idea."

"Sergeant, notify the rescue ship's captain that I accept his very good collision course idea. But after two hours of approaching each other on collision courses, flares must be used by both vessels at five-minute intervals. I will greatly appreciate his using the three-inch deck gun at five-minute intervals also, until we make contact. Thank him for his help and inform him that we are in a snow fog with very limited visibility at best, but we will begin flares in two hours and listen carefully for our mother-ship's deck gun."

"Will do, Captain."

Sam asked me if I planned to use our quick one-eighty course change strategy.

"Yes, are you ready?"

"I'm ready."

"All hands stand by to come about!" I shouted. "The bow's heavy with sea water, Sam."

"Okay, I'll remember that."

"I'm on an excellent wave now, Sam. EXECUTE! Perfect, even in a fog! Sergeant we need a conference as soon as you can break free!"

"Be right there Captain."

He approached, assuring me that the collision problem was under control.

Chapter 48

"As you know Sergeant, in about two hours we must be hoisted aboard our mother ship, even in this worsening snow storm, or we will not survive. But there are two things that give me great concern."

"What's that, Captain?"

"First, all of our collision damage seems to be, so far as I can tell, in and about the Samson post, which is the primary source of my concern. If it is damaged, we cannot be hoisted aboard our ship because that's where the forward fall and the sea painter attachments for hoisting are located. If either of them fails during the pick-up we are dead people! Second, what I want you to do is inspect the damaged area and report back to me with your opinion about hoisting this heavily damaged vessel aboard our mother ship."

"Captain I really appreciate your confidence in me but I believe you are better qualified to do that than I am."

"Well you are probably right but there are many other things you do better than I, so let's do it together. Hawkeye, take the tiller while the Sergeant and I go forward to evaluate our boat's collision damage, our course is northeast."

"Aye, aye Captain."

Upon our arrival at the point of impact, I immediately detected that the water was slowly gaining on us, just as I had feared. I ordered Guns, "Stand by with Sam's super bilge pump, and keep the boat as dry as possible until we are hopefully hoisted aboard our mother ship. If the time comes that we will be hoisted aboard, I do not want one drop of water in our bilge because extra weight at that time we do not need. I realize some of our weight is snow but remember, snow weighs over eight pounds per gallon when thawed, the same as water, and it will get ahead of you before you can turn around. Do you men have a battle lantern up here?" I asked.

"No Captain."

I walked aft to the engine spaces, got a lantern from Sam, and gave it to Guns. I advised him to use it to monitor our flooding condition and above all, to keep me informed.

"I will do that Captain."

The Sergeant and I did everything possible to test the soundness of the Samson post and the sea painter cleat. Then by mutual agreement we decided to go for broke and accept being hoisted aboard the rescue ship because that was the better of two very deadly alternatives. I relieved Hawkeye at the tiller so he could begin shooting our flares, straight up, at the agreed-upon intervals. It was a little early to begin but I felt very uneasy about being on a collision course with any vessel, and especially after remembering that, in this storm, sound from a deck gun and light from flares would be of limited anti-collision value. Therefore I ordered a course change to five degrees west of our agreed-upon heading and put everyone available on lookout duty.

It seemed as if we would never hear or see any signal from the rescue ship but finally it happened! I ordered "heave-to" (stop) because I did not trust two vessels underway approaching each other during restricted visibility conditions. Soon, some of us barely heard the deck gunshot and with considerable uncertainty also reported a flare. After that, things began to happen fast! All attempts from both vessels to contact each other were changed from five-minute intervals to one-minute intervals and wireless messages, which were instantly shared with me, had begun to totally occupy the Sergeant's time.

We were all very thankful for the probability of being rescued, but without a doubt we were about to begin the most demanding and dangerous boating experience of my entire Navy career. It took no less than an hour just to maneuver both vessels into favorable positions for a possible pick-up. Getting a badly damaged, sinking boat on board a ship, that was itself bouncing all over the place during a treacherous winter blizzard, offered little chance of success, but it was the only chance we had.

Our mother ship's captain stayed in contact with me using the Sergeant's wireless. All other means of communication had been snuffed out by the storm's cyclonic churning of the atmosphere. And without good and timely communication with him we had no chance of success at all. Our rescue ship had enough engine capacity to run with the storm and give us at least some lee side protection, without which pulling us aboard our rescuer would have been absolutely impossible. But it took Sam and me considerable time and skill to get under that wing of hope (lee side).

Sam asked, "What's your plan, Captain?"

"Can you get me close enough to retrieve the painter if they use their messenger line?"

"Yes, I'm sure that I can!"

"Do you have radio contact with the pick-up crew?"

"Yes!"

"Good! I'll keep us about ten feet clear of the hull until we attach the painter."

"Captain, I'm getting low on fuel. We can't do this a second time!"

"That being the case, my friend, there will be no second time because we'll get it right the first time."

"That's exactly what I needed to hear from you, Captain. Let's show them what we can do when the odds are against us!"

"Well said, Sam!"

"Ready Sergeant!"

"The best assault team in the Navy is ready, Captain!"

"Okay, have two men, one aft and one forward, to hold a boathook upright as high as they possibly can. Then instruct the pick-up crew to have their strongest man to throw a lead-weighted messenger line over us, not to us, and pray it falls between those upright boathooks because we have dry fuel tanks which means no second chance. I'm also sure you realize that no engine means no bilge pump."

"I'll go forward, Captain, and make sure the painter is attached as soon as I get my hands on it."

"We'll need every second, Sergeant!"

Chapter 49

Every team member was firmly reminded to be on the lookout for the thrown small-gauge messenger line, and that whoever retrieved that life-saver should immediately pass it forward to the Sergeant. Thanks be to Almighty God that someone quickly shouted "GOT IT, I'VE GOT IT!"

In only a few seconds the Sergeant shouted back to me, "The sea painter is secure on its cleat and it's holding!"

"Put her under the hoisting falls, Sam!"

"Okay Captain but in this wild sea they will need to have plenty of slack, because those heavy hoisting blocks will probably be in the boat most of the time!"

"Yes, I believe you are right, Sam. We must have a lightning-quick hook up and hoist to prevent their bouncing around in the boat too long, which could possibly rupture our hull. Using my headphones I've already informed the boat retrieving officer about that problem, and have also requested that deck lights be turned on after the sea painter has been secured, even though it's still early p.m. Because in this storm, visibility is very changeable, and we do not know if or how long it will take for us to be hauled out of the water. And now Sam, seamanship must take precedence over all unnecessary talk."

"Aye, aye Captain!"

"Attach the forward fall, Sergeant!"

"Forward fall attached and ready, Captain!"

"Ease her back a bit, Sam, and test the Samson post towing strength!"

"It seems to be holding okay, Captain."

"Good! Now put us on full tow."

We're on full tow now, Captain, and all seems to be well!"

"Attach the aft fall, Hawkeye, and then stand by to remove the bilge drain plug the instant we start moving up, because we need to get rid of excess water weight as quickly as possible."

"Aye, aye Captain!"

I wrote about monkey lines earlier, but perhaps further explanation at this time would be helpful to the reader. A steel cable is bolted to the outboard end of each davit (boat support) so that when they are fully deployed, the ship's hoisting equipment will be precisely over the hoisting attachments on the boat. To that cable several ten-foot safety lines are firmly secured to insure that if the boat should have a mishap, passengers' lives could be saved by their grabbing and holding on for dear life to those dangling ropes, which are nicknamed "monkey lines."

Seconds after we began being hoisted up, our engine sputtered and died because there was no fuel left! Consequently, there was no bilge pump left either, and seawater quickly began filling the boat and thereby greatly increasing its hoisting weight, which as captain I decided was the end of our boat's recovery attempt.

"Everybody grab a monkey line and hold on for your life," I ordered, "because our boat is about to break apart and sink." Our mother ship's captain became aware of our crisis and slowed the ship as much as he could, considering the blizzard. Which wasn't much! After we had been hanging free on the monkey lines for only a few seconds, large swells began to slam us again and again against the rescue vessel's steel hull, but we all by the grace of God and our survival instincts managed to hold on until welcomed hands from heaven (rescue people) dragged us up onto our mother ship's frozen main deck. From there the medics took over.

However, one of the Rangers refused to release his grip on his monkey line, so the medics cut it and took him to sickbay while still holding onto the remains of the cut line. There were many good-natured laughs about that later on.

I was, according to my medical records and the ship's doctor, borderline qualified to have been on such a mission in the first place. But now he said, "There are new problems that are beyond our medical capabilities. Tag him emergency and get him to *Dixie* medical as soon as possible! In the meantime, put the Sergeant who was earlier taking care of him back on his case, if he (Sergeant) survived the mission."

On the main deck, the determined retrieval crew raised the bow of my boat to empty some of the excess water weight, and then brought the stern to level and placed her in a steel cradle on the boat deck. I was thrilled when the Sergeant told me that. There's just something more than memories about a boat in which one has endured so many life-threatening experiences, and that special something is quite difficult to write about in words that others would understand. But the only way I know to share them is to just express my feelings. I loved that battered old war horse and I will mentally carry her with me until my final breath. I don't believe she will ever put out to sea again. But the admiral will, after remarkable service, have his engine returned.

The storm passed the following day and I was evacuated by helicopter to USS *Dixie* medical. However, leaving my teammates was not an easy thing to do. I barely had time to say farewell to brave men who, unknown to me at that time, I would never see again! You probably will never believe it but my doctor on board *Dixie* was another female Navy surgeon. Female suited me just fine, but why a surgeon? Now that title frightened me a bunch!

There were several medical people around a table, upon which I had been stripped naked and laid out. They were debating the pros and cons of amputating my right arm. I was in sort of a twilight zone but I could hear what was being said. However, I understood next to nothing about their medical talk. My case doctor, a Navy commander, said in a rather angry tone, "This sailor was sent back north before that arm was healed enough to handle such an extremely cold environment. I recall the hurry-up debate. Just look at this arm! Frozen again! I'll need help on this one bigtime!"

Chapter 50

"May God help me!" She was called away from my table and left the others preparing her patient for whatever was being considered for my medical needs, which at the time had done nothing but fill me with fear of losing my right arm. She soon returned and said in a whisper, "This is him! He's all over the news about something he did in Russia. Prepare him quickly and move him to isolation. I'm waiting for a phone call from captain [deleted], she is working on a case at San Diego Naval Hospital. I'll join you as soon as I can."

Suddenly, it seemed that keeping me from talking to anyone became more important to somebody than did my medical needs. I thought of my teammates who would also be debriefed and go through the same thing, only this time it would be much worse. We were told to stop those damn floating mines and that is exactly what we did! So what's all the fuss is about, I thought.

The fuss came from Russia at the UN who while raging before that body was accusing the USA of attacking their country without cause. Of course there was no mention of the fact that they (Russia) were sinking our ships with made-in-Russia floating mines. Also there has never been any hard evidence that our assault team was responsible for those two Yalu river attacks. Just before I was moved to isolation, I requested from my doctor that my survival suit be brought to me.

"I'll get it. Where are your clothes?"

A nurse who had stripped me naked reported that it was necessary to cut my clothes before removing them to prevent further injury to my arm.

"That's all right, ma'am. Please get the survival suit for me because there is something very deadly hidden in there that I must recover."

The doctor handed them to me and said, "What could you possibly want with this bloody mess!"

The rough experience on the monkey line had caused the very thin skin on my injured arm to begin bleeding again, and that considerably disturbed my doctor who spoke up and vowed that my career as an assault coxswain, as far as she was concerned, was over! I fumbled around in my foul-smelling clothing until I located the small packet that contained my cyanide pill, which I passed on to her (doctor) and said, "This is my suicide pill. I do not want this dangerous poison to get into the wrong hands, but anyway I don't need it now – do I?"

Everyone who heard that got a big laugh. A man dressed in civilian clothes appeared out of nowhere and rudely said to my doctor, "I'll take that!" She then asked if he wanted my other clothing that had been removed. "Yes!" he curtly answered.

I truly believe there is a special training facility hidden away somewhere in our military that teaches bird-brained men like the one in my quarters to be discourteous to women and thereby feed their sick personal image of superiority. It has always been very uncomfortable for me when I hear men who take pleasure from publicly being hateful to women, and I, with no little delay, gave him a good verbal dose of my thoughts on that subject, in the presence of all of my caretakers. My doctor eased up to my side, gently put her hand over my mouth and quietly said, "Hush hush sailor boy, you have said enough." I knew the routine well because I had been there before and there was no cause for Napoleonic bad manners toward those who were so deeply involved with my well-being. I was rushed away under guard to isolation where medical care for me was continued.

But before I was put to sleep, the commander placed a phone to my ear and said, "Captain [deleted] wishes to speak with you."

We exchanged friendly greetings in which she told me how happy she was that I had survived such an awesome mission. "I'm so proud of you but presently I cannot even acknowledge that I'm your doctor because the commies are determined to capture or kill you because of that last attack! And I do not want to help them find you so I will fade into the background. However, for the present time commander [deleted] is your on-site surgeon and I have total confidence in her ability to care for you. We have had a lengthy telephone conference about your case, and believe me when I say that you are very fortunate to have her as your case manager. But trust me, I will keep in touch with her! May God be with you and I must say that your having been a part of my life is something I will never forget."

"Thank you ma'am, I share those feelings and your call has given me the confidence that I need. Maybe we will meet again sometime and hopefully I will be able to hug you with both arms. But for now there is one last thing I must say, I will never forget you and the fact that you made it possible for me to achieve on my final mission what I so badly wanted, and that was to protect my Navy brothers! Goodbye and best of luck always."

"Now," the commander said, "it is time for you and I to have a talk. It is necessary that I put you to sleep for an exploratory examination of your injury. I expect it will take two hours or more. The history of your injury leaves me absolutely no room for error. I must get it right or you will surely lose your right arm. Therefore, I must know exactly what and where I'm treating you before I begin."

"Do you think I'll lose my arm?"

"That, I cannot say at this time, sailor boy, but I can tell you this, I will save your life."

"When will you start, ma'am?"

"Right now!"

"That's okay with me. This arm has been hurting me long enough. However, I'm not prepared to give it up just yet because I need it along with the other one for hugging!" That one brought the house down.

Chapter 51

I was put to sleep while hearing the nurses talk, they were saying things like "Hang in there sailor you owe me a hug!" Two hours became, unknown to me, four hours. Vaguely it became apparent they were trying to awake me, which I was later told took over an hour. It took another eight to ten hours before my mind was clear enough to communicate sensibly with the commander.

However, when that time came, she spoke the words I had prayed for. "Young man, I'm happy to report that you will keep your arm!" I became emotionally overcome and could not speak for several minutes. However, after I had regained control, thank you's a dozen times over were said to all those who had made it happen – not forgetting God Almighty.

"Now you must remember there will be rules and you must do as I order, or I might need to go back in there. But if that should happen I do not believe I could ever again reach the level of proficiency that I experienced today, because God was with me in a very special way. This has been a long and complicated procedure so don't mess it up! And there is another thing you can count on, you will not be re-certified by me as an assault boat coxswain!

"My assistants will take over now. I'll see you later today!"

"Thank you commander."

Before she left I took notice that she was talking to my Marine guards, who promptly escorted the civilian from my quarters. The following day the arm that had been bandaged to my belly had very little pain and that was a blessing I had long hoped for. The nurses sat me up on my bed and helped with a grand breakfast. Next they asked me to stand, which with only a little help I was able to do. Throughout that day they were constantly attending me. They must have been following a recovery schedule or something of the kind. Anyway I was exhausted at the end of the day.

Events moved along in like fashion for five days before the commander would permit the civilian to interview (interrogate) me. It was my understanding that she had a considerable fight over that issue. Her kicker was that I was still under the influence of drugs. Anyway the time finally came as we all knew it would.

"Good morning young man my name is [deleted], I represent ambassador [deleted]. Do you feel well enough to talk with me for a few minutes?"

"Yes sir, I do."

"Let me begin by saying that what you and your team did is absolutely the most heroic thing I have heard about in this war! But I guess you are aware that you and your team cannot at this time be decorated for your brave service to our country because we (USA) are hotly debating Russia at the UN right now claiming that we know nothing about those river attacks." (I was decorated for that service fifty years later). "My primary purpose for this interview is to learn facts about your recent mission to Russia, so I will begin by asking a few questions, if that's all right with you."

"Yes, I hope I can help you."

"Thank you. At what time did the attack occur?"

"Well sir, I had planned for a dawn attack but it was snowing very hard and that considerably restricted our visibility. So I held back until I could actually see the Sampans (targets) I had planned to attack."

"Can you pin down an approximate time for me please?"

"My guess would be about seven a.m."

"The next question is very critical for supporting our debate with the Russians! Do you believe that your vessel could have been identified by anyone during your time in Russia?"

"No sir I do not. Because if they had seen us, they would have tried to either kill or capture us."

"Can you support that claim?"

"Yes I believe that I can."

"Please do so."

"Well, I'm sitting here talking with you and that alone should be evidence enough."

He laughed and said "Good comeback Captain but can you elaborate a bit more?"

"Yes sir, if Russia had any evidence, there would be no debate, because that evidence would be posted all over the world. My team is their only hope for evidence and that is why they so badly want to capture us."

"You are right about that, Captain. Therefore, we can not allow that to happen."

USS *Perkins*

Image: Chuck Snyder

"It may be helpful to know that there were close to a hundred Sampans loaded with Russian mines anchored in the river delta, waiting to put out to sea on the morning tide, using a head-on course toward our fleet."

"How did you know they were loaded with mines?"

"Before we fired a shot it was clear to us that those Sampans were not rigged for ocean fishing because it was too cold for that activity. And there were a large number of them that had anchored very far apart with no anchor lights, which was also a good indicator that they were loaded with mines and were protecting themselves against an accidental explosion. Every one of them had clear signs of being mine bandits. I did not see anyone on any of those Sampans but they were surely manned and loaded with mines, or they would not have been anchored that far apart and that far away from the shipping channel, which they clearly were.

"My assumption was that since it was so bitterly cold, our enemy had bundled up with everything he had to keep from freezing to death, and consequently could not hear our engine as we approached them. You can be sure that either my team's action or the freezing water killed them all, and of that I am positive. Therefore, it goes without saying that we left no witnesses."

"What did you do after that fight?"

"Our decision to commit suicide came next. After our having created so much noise from the delta mine explosions, we expected Russian helicopter pilots would guess our escape route to be east on the Sea of Japan. However, we decided to go up-river instead and fight until our last breath."

Chapter 52

"We all knew that during daylight hours with a visibility range of about one thousand yards and improving, the enemy could blow us out of the water with very little effort because there was no place for us to hide. So we began motoring upriver (entrapment) as fast as we could, on what every team member believed would be their final boat ride. We also believed the ice barrier on the northwest or Russian side of the river would be the most likely place to search for the grand prize – the staging area where our enemy was loading two to three floating mines onto each Sampan. That was the target we had prayed for.

"After twenty to thirty minutes of motoring up-river, we still had not encountered any resistance. The Sergeant and I were completely puzzled and he ordered every man to their battle station, expecting the fight to begin at any moment. We sped on to the ice barrier (ten miles upriver from the entrance light) where we, still unopposed, discovered the target of a lifetime. The Sergeant quickly led his Rangers ashore and began the attack by attaching time delay (15 minutes) explosives to the mine storage buildings that cumulatively housed thousands of mines waiting to be loaded onto Sampans. But before the time delay bombs exploded, the Rangers destroyed hundreds of Sampans and many other targets which were necessary for the Sampan mine operation – all without any resistance from anyone. We were completely unchallenged!"

"No enemy fired a weapon at any time, even from a hidden location?"

"Absolutely no one fired a weapon at us at any time while we were destroying Russian targets in their country!"

"I must ask you again, are you SURE that no one saw any of your team while they were on Russian soil?"

"Yes sir, of that I am quite sure! Please be reminded sir that an assault boat captain is always the first to be taken out to prevent opportunity for an escape. Therefore, I know better than anyone that there were no bullets fired at me and neither did any bullets strike my boat."

"All right, now let's move on and review what has been said here today. And remember, you may make changes if you wish or even withdraw an answer because you are not sure about the facts."

"Thank you sir, I'll remember that."

At the end of the interview, he was exceedingly pleased that I had stood firmly by the fact that none of us had been seen on Russian soil. Just before he left he asked if I had ever heard of Asian flu.

"No sir, I haven't."

"It's an epidemic disease that has killed thousands of people in east China and Russia, that's why you didn't see people in those areas you spoke about, they were afraid to be around others for fear of becoming infected."

"Well sir, the Lord sure does His work in mysterious ways."

"Yes he does Captain! What you have told me here today will sure make our Ambassador happy because the Russians are claiming they have proof that your team violated international law by illegally entering their country to commit acts of war. But good chemistry has quickly developed between us and I believe every word that you have spoken here today."

"Thank you sir!"

"Captain, before I leave I want to thank you for reminding me that I should always be courteous to women, I will in the future try real hard to remember that and govern myself accordingly."

We both smiled as he walked through the hatch (door). The majority of my debrief was over and I was pleased about that. There were several clarifications requested as time passed, but I was always glad to help my country any way that I could.

The closely-managed healing process took six weeks. After which I was allowed to, with my two Marine guards, move about *Dixie*'s main deck one hour per day. But I was denied the luxury of talking to anyone.

I didn't count the days, but perhaps on the tenth day during my walking time I saw a beautiful sight. USS *Perkins* DD877 was smartly entering Sasebo harbor. I requested from the Marines a few more minutes to look. At almost the same moment my doctor, the commander, who was jogging on the main deck, approached me and asked how things were going. After answering I called her attention to my ship entering port.

"Isn't that a beautiful sight commander?"

"Indeed it is Captain!"

We continued watching while her captain maneuvered his ship to a port-to-starboard nest-docking with three other destroyers and their mother ship, *Dixie*.

"When do you see me again?"

Passing some time on the stern of the "*Perkins*".

Image: William C. Warwick, Jr.

"Tomorrow at ten a.m."

"That's good because I have an idea that might interest you."

By the next morning I was exploding with curiosity about her idea. I arrived for my appointment right on time because I was anxious to hear what she had to say. Social graces were quickly expressed and she quickly began explaining the reason for our meeting. "Word has leaked out that you are on board *Dixie*. Therefore, the civilians have requested that I release you as my patient. I will do that after this meeting. If you agree, I will use whatever influence I have to get you on board *Perkins* since you will soon be discharged as one of her crew."

"Ma'am that would please more than anything!"

"She will be nested with *Dixie* and others for five days and during that time you could easily be smuggled aboard. How does that sound to you?"

"Absolutely wonderful!"

"Good, I will assign a code name to your records so I can still doctor you until you are discharged. Be prepared to move at any time and I hope you will go to *Perkins*."

I stood, saluted and asked her permission to pay a debt I owed. She stood and answered, "Permission granted, sailor!"

That appreciative hug forever ended my acquaintance with a second woman to whom I am indebted for still having a right arm. So if you happen to be someone who takes pleasure from being cruel to women, I will be your worst enemy! Goodbye ma'am I'll never forget you!

Chapter 53

The Marine guards (actually protectors) rushed me back to my quarters, where what few things I owned had already been packed, and were ready for me to be moved by the civilian agents to someplace that would be safer. I was told that the so-called UN debate had turned into an ugly shouting match. They arrived to move me soon after I had returned to my quarters and told me that this time they would use a look-alike to draw the Russians attention away from me.

My protectors believed that the soviets had fully directed their attention toward *Dixie* because they believed I was on board that vessel. Other agents would later move me off *Dixie* while the commies were looking the other way. KGB was closing in and likely had the pictures from the fish factory, one of which was taken of my being helped to the rescue helicopter because I was unable to walk rapidly. And another thing – people came and went from that large ship (sixteen thousand tons) day and night. Therefore, it is very likely that KGB could put an agent on board *Dixie* anytime they needed to do it, because capturing me was a very high priority and would certainly put a star in some Russian agent's crown.

Few people realize that the communist party in Japan was very large during the Korean War because everyone was not happy that Japan had lost WW II, especially to the USA. It might be interesting for the reader to learn that *Dixie* was at that time docked in Sasebo, Japan at the end of its main street, which has a long, gentle, up-hill climb from the water to an elevation of approximately one hundred feet. Her entire starboard side offered a perfect sniper target. And guess what, one could disembark or embark from *Dixie* only by using her starboard gangway!

Now, it goes without saying that I was not a welcome guest and the crew very much wanted to move me before some unlucky soul got killed due to my presence on board their ship and my mother ship! I was not told anything about the unfortunate look-alike or anything else except to stand by to be moved. My first move that day was through *Dixie*'s deepest internal spaces where no one but specially trained shipfitters could navigate. It took considerable time but we finally came out of the darkness into the light of the machine shop where I was disguised as a welder's striker (helper), and my next move was carrying welding equipment to *Perkins*.

When I arrived below deck out of sight, my shipmates loudly welcomed me home and took good care of me until we got underway. To do that they locked me in the aft steering compartment, where absolutely no one could gain admittance except those who had been assigned to that watch station. Shortly after *Perkins* had cleared the sea buoy on her way to the bomb line (area most affected by floating mines), my shipmates brought me topside into the cold sunlight, and after that to the bridge deck where I was welcomed home by my captain.

After a brief conversation about my physical condition, he ordered, "Feed this man and find him a good place to sleep because I'll need him before this cruise is over. You are dismissed gentlemen!"

We all saluted our much-respected captain and left the bridge. I was assigned by the chief boatswain's mate to some of my previous duty stations except the MWB! However, time and circumstance would soon change that order. All of my shipmates somehow knew that I had been restricted to limited duty and they made sure that I did not disobey that order. Consequently, I felt as if I were a tourist, or at times in their way. The only thing left for me to do was to walk about my ship and watch my friends busily prepare for their next battle, which was just around the corner.

When I first came aboard *Perkins* in nineteen forty-eight, I did many of those same things but now after two wars I'm told that I must stand aside and let others do things I was forbidden to do. Not a good feeling, no not a good feeling at all. However, I would soon be discharged and war in my life would forever be put behind me – or so I thought. The sad truth is that it has never left me, no matter how hard I have tried to purge it from my memory. War is never over for those who have participated in it. While at sea, I spent most of my time on the boat deck dealing with endless boredom. There was no one who had time to waste sitting with me just to chat.

Suddenly there was a lot of commotion at the MWB. Definitely there was something unexpected happening. The shipfitters who had been working on her had stopped as if they were very upset about whatever it was. I also became very concerned but remained seated, and resolved to stay out of whatever it was that was bothering them. Then I became aware that all of them were looking at me. I remained seated and thought, If there is anything that concerns me, I will know soon enough. Since they did not approach me, I remained seated.

The chief came up on the boat deck and approached the idle shipfitters and they all appeared to be having a rather lively conversation. After a few minutes had passed, the chief left, waved to me, and climbed to the bridge deck. Something big has happened which the captain should know about, I thought. I had chosen a very good place to sit and observe that there was a problem. But what? Next two Army people appeared and climbed up to the bridge. Well, whatever the problem, it did not concern me so I refocused my attention to other things. How wrong I was!

Chapter 54

Perhaps the most dangerous mission during my final days in Korea was one in which I carelessly drifted into being a "show-off," a state of mind from which the consequences were near to being fatal. Since *Perkins*, nicknamed "hell ship" by our enemy, had become well known as a very aggressive close-in fighting destroyer that had achieved an extraordinary combat record, two reporters from "Stars and Stripes," a military news publication, had been allowed to come aboard for a close-up observation of her crew's much talked about fighting reputation. They also wanted to interview the assault coxswain (me) who had been nicknamed "Wildgoose."

The captain refused to allow any reference to my name to be used and clearly explained to them why I had recently returned to his command! Subsequently, they asked our captain's permission to go along with me on a target spotting/hunting mission. He referred them to the medics who had placed me on limited duty while waiting for transportation to the separation center in San Diego, California. I had spent most of the morning hours watching while the shipfitters were preparing *Perkins'* motor whale boat for the upcoming night mission. Unexpectedly I was paged on the ship's PA system and ordered to report to sickbay. What now, I thought.

Upon my arrival I was introduced to the Army reporters and then told by the chief medic the reason for their being onboard. "Boats (me), I have a problem and you also have a problem! My problem is that I have confined our regular assault coxswain to his bunk with an unknown illness and the captain will be very put out with me unless I release you for tonight's mission."

"Yes, I'll do it!"

"Wait now, not so fast, I haven't released you yet!"

"I know that but I hope you will."

The chief asked, "Are you reporters sure you want to go to war with Wildgoose?"

"From what we have heard about him, yes we do."

"All right, I'll inform the captain. Boats you must remember this, I will personally escort you to the boat before it is launched, to make sure that your right arm is warmly bandaged before you leave. And also you will wear warm clothing which includes gloves. Agreed?"

"Yes, I agree Chief."

We left *Perkins* several hours after the noon meal and motored to a small offshore island which I had used in the past to seek out possible targets, some confirmed some not, on the mountainous mainland. My radioman received a late afternoon message for me to intercept and destroy two floating mines. I was informed that an out-of-ammunition Navy aircraft had discovered and reported their exact location, but because of being unable to shoot, could not destroy them. However, the aircraft would remain over the targets, if possible, until my arrival.

From the compass bearing I had received, the two dreaded hazards were located about five miles offshore, beneath the circling airplane. The pilot sighted me, rocked the plane's wings, and returned to his carrier. I was no stranger to blowing up Russian floating mines, which were not uncommon off the East coast of Korea; I had done it more than a few times. After evaluating sea conditions, I put my boat in the most favorable attack profile and ordered the engineer to give me a high speed run toward the two navigation nightmares.

It was show-off time! With an M-1 rifle I locked and loaded nine rounds (bullets), cautioned everyone to keep their head down, and steered a direct course toward the targets. After firing four rounds at the nearest mine, it exploded with a horrendous sound, sending water and black smoke 200 feet up into the sky. With little course change or delay, I sped toward number two and began firing at it. The frightened reporters shouted, "Slow down Wildgoose, you are getting too close, you could kill us all!"

But I had already begun firing at number two, which took five shots to destroy. By that time we **WERE** too close and the boat was caught up in a huge swell which was created by an almost submerged explosion. That grandstanding foolishness almost broached (turn upside down) the boat, and no one on board was at all quick to forgive me for their ringing ears, caused by being too close to the mine's deafening final moment.

Previously, I had always required everyone on board to use ear protection. Purposely, I had tried to frighten the reporters and consequently came dangerously close to losing my boat and all in her. No more showing off, I silently vowed. After reporting that we had successfully destroyed both mines, I asked and received permission to return to my island. Several hours later we would, as earlier planned, direct *Perkins*' main battery (six 5-inch 38 caliber guns) at an enemy military supply train which friendlies had advised would be traveling South through our area at about midnight. All trains in North Korea were coastal because laying tracks in the mountainous interior was too expensive.

We had plenty of time to prepare and I decided that the train's steam engine must be targeted first because after it had been stopped the rest would be easy. I directed my engineer to tie our boat's battle lantern to a carefully selected tree near the RR tracks and then rig a trip line so the engine's forward movement would turn on the lantern at a perfect place to shoot. *Perkins*' range and bearing to the target area had to be guesstimated. However, after firing a marker shot, I could by radio put the following shots where I wanted them to be, assuming we (my team) had not been killed by the marker shot.

We heard the train coming but could not see it. I radioed *Perkins*' gunners to stand by to shoot. When the lantern light came on, I shouted twice into the radio transmitter "COMMENCE FIRING!" The first shots exploded about three hundred yards south of the battle lantern light. "On target … keep 'em coming." I had lied to the shooters because it was my guess that the southbound engine would very soon be exactly where I wanted it to be. I wasn't in the business of conserving ammunition so I called for continuous firing.

Chapter 55

The engine kept moving, just as I had expected, until it was located directly beneath my shots. Consequently it was quickly taken out. Confirmation of the kill became evident by the widely scattered burning coal from the engine's firebox, and lots of derailing sounds, accompanied by a boiler explosion. We got it, I thought. Then as I further directed hell ship's shots north along imagined RR tracks that I could not see, the train's cargo (ammunition) began to continuously explode for ten minutes or more with very loud noises. When there were no more explosions, I radioed hell ship's gunners to "CEASE FIRE, mission accomplished."

Then I decided it was best to get out of there FAST before the enemy became overly curious about the revealing sounds of high caliber naval gun fire and many other explosions that could not be as easily identified … at night. The reporters were obviously very happy when their close-up war experiences with me and my team were over. However, those desk boys had, on that day, witnessed real war with an enterprising U.S. Navy assault boat crew, which had included several close-up combat adventures that ended for them only after our boat was safely back in her steel cradle onboard *Perkins*, which was about 4 a.m. They wrote a flattering article about hell ship's gunners and their wild (their word) assault boat coxswain.

Our captain allowed nothing to be published that would identify me as the same coxswain who the KGB (Russian Secret Service) so badly wanted to capture. The chief medic was, as per his request, notified as soon as we had

A typical Sampan.
Image: William C. Warwick, Jr.

attached *Perkins'* sea painter to our boat's Samson post. Immediately after I disembarked from the boat, he requested that I accompany him to sickbay so that he could rest assured that my right arm injury had not been aggravated. It had not. Therefore he was well pleased and it goes without saying, so was I!

"The captain is not yet through with you, Boats. I have been ordered to report to him in his sea cabin as soon as I can confirm that you have safely returned, and that your physical condition is none the less than it was before you left. I have arranged for *Perkins'* regular assault coxswain to be air-evacuated to *Dixie* as soon as possible because there is nothing more I can do for him on board *Perkins*. I'm sure the captain will ask me to release you from limited duty for the duration of your time under his command. Then you will receive orders to report to the separation center in San Diego for discharge. How do you feel about that?"

"You know how I feel, Doc! Sitting idle while my shipmates are doing the fighting, that's just not my cup of tea."

"How well I know that, Boats, but I'm caught in the middle of two high authorities. I'm damned if I do and damned if I don't!"

"Chief, I have an idea that might well serve both of us."

"Okay, let me hear it!"

"Let your records show that I report to sickbay daily for evaluation, which I will do before each mission, so that much of it will be the truth. Anyway, these short night missions compared to the frozen north for a week or more are to me nothing more than limited duty anyway. Next, we both know how desperately the captain needs me. Therefore, let's get him involved. Let my medical record show that you released me from limited duty based upon the captain's authority. That also is the truth. If my memory serves me correctly, a commanding officer has the authority to use his people (crew) as he sees fit, especially in a war zone."

"Good idea, Boats, that is exactly what I will do. However, you must continue to do as I say and stay out of unnecessary trouble."

"In my business that is not always easy to do but I will promise you this … I will try."

"Thank you Boats."

Before my resting period had expired I was awakened and alerted that Sampans had been spotted by a Navy patrol aircraft. "So far eight have been reported and they are launching mines."

"Tell the chief (boatswain's mate) that I'll need three gunner's mates armed with two BARs, one M-1 and plenty of ammunition. Also ask the cook if he will make a few quick sandwiches for us because this is probably going to be a long day!"

I raced to sickbay so that "doc" could prepare me for what I believed would be a tough fight with eight or more Sampans. He made sure that I was warmly dressed, including gloves, and reminded me of my promise not to do anything foolish. We rushed to the boat, which had already been lowered to the main deck railing. I jumped aboard and called for system readiness reports.

"Engineer."

"All systems ready Captain."

"Fuel situation?"

"Fuel tank is topped off, Captain!"

"Gunners."

"Three gunner's mates are armed and onboard, Captain."

"Weapons."

"Two BARs, one M-1 rifle, seven sidearms, and plenty of ammunition."

"Captain's talker (radioman)."

"All communication is working properly and synchronized with *Perkins'* bridge deck captain."

"Sea painter."

"Attached and taut, Captain."

"Bilge plug."

"Screwed in tightly and tested Captain."

"Forward fall."

"Ready to lower away, Captain."

"Aft fall."

"Ready to lower away, Captain."

"Tiller lock pin."

"Secure and ready, Captain."

"Mates are we ready to go to war?"

With a moderate volume my crew answered, "Let's go get 'em Captain."

"Engineer start the engine and tell me when you are ready for us to be lowered onto the sea."

"Aye, aye Captain."

During that very brief wait, the cook passed several containers of food to us.

Finally the engineer reported to me that "The engine is running perfectly and ready to be lowered onto the sea."

"Good luck men," the chief said. "Stand by to lower away fore and aft. Together now, lower away easy, that's it, easy now, slack the sea painter," the launch officer directed.

This time it is a piece of cake: five knots and a peaceful sea, I thought. Using the navigation aids which had been given to me by *Perkins'* navigator, it took close to two hours to locate our target. But when we did, the prize was much larger than we had anticipated – it was a joy to behold!

Chapter 56

My binoculars revealed fourteen Sampans, and mines all over the place. If we totally destroy this effort to sink our Navy's ships, lives of many sailors will be saved. My observation was that the mines were drifting south toward us. Therefore my attack plan must take that danger into serious consideration, I thought.

"Engineer, throttle back enough so that everyone can hear me. Mates, as you can see, the mines that are already on the water are moving toward us on a considerable current. Talker, report to our captain that we have located the targets, and many mines are drifting due south with an estimated two-knot current. Our position is longitude 135 east, latitude 40 north.

"Aye, aye Captain."

"Gentlemen, I have decided that it will be best to attack our enemy from their north side and avoid until later the unpredictable drifting mines on their south side. I will steer toward the target that I want you to take out first. You must not destroy either a Sampan or a mine that is located less than five hundred yards away from us, because if you do, we could be blown up with your target or be killed by its shrapnel. I will give you plenty of time and plenty of shooting opportunities but the decision to shoot will be yours. There should not be any confusion if the shooter disciplines himself to his intended target and only that. Remember, I will make as many passes as necessary, so don't be impatient! I want the BAR gunners at the port and starboard bow and the M-1 gunner in the cockpit with me to protect my back while I'm zigzagging all over the place choosing targets. Don't forget the ear protection that I have given you because there will be a lot of noise this day!

"Now one last thing, if any of you have personal reservations about killing or helping to kill thirty people or more who are trying to kill us and our shipmates, tell me and I will excuse you from this fight right now. However, I will not excuse anyone who fails to do his duty after the fight has begun and remember that on this boat I am your commanding officer. And if it should become necessary, I will charge anyone who holds back and will not fight, with refusing to obey my orders in the heat of battle. Let everyone be assured before this fight begins, that I will not quit until the last enemy has been killed, the last Sampan has been sunk, the last mine has been destroyed, or until I have been killed. Have I made myself clear?"

"Yes Captain you have, let's go get 'em!"

"Engine full ahead," I ordered. "Gunners, stand by to shoot at will after I have given them a target."

All fourteen Sampans were separated horizontally, east to west in a calm sea over about five to six miles, which made them an easy target with little resistance. But their chief reason for their wide separation was to enlarge the drift pattern, and also for their protection from an accidental explosion. As I made my turn to begin attacking the enemy's north side, I heard the haunting sounds of weapons being prepared to shoot. God be with us, I prayed as I steered east to west at a safe distance from anticipated explosions (excepting the few seconds it would take to point my bow toward the Sampan I had chosen to be destroyed first).

After the first one had been successful, I quickly returned to my estimated safety zone and repeated the same procedure until all fourteen of the bandit Sampans had been reduced to splinters (the wood would not sink). There were undoubtedly many mines that still remained on board the Sampans and consequently were exploded with our gun fire. The only enemy return fire we received was more than likely coming from WW I-era .31 caliber rifles, which had an effective range of only three hundred yards, and that would be on a good day – obsolete gifts from Russia.

All of their shots at us fell short of the mark because I stayed clear of the mine explosions. My estimation was that twenty mines had already been put on the water before we arrived and that made for a long, dangerous job for us because we had to destroy them one by one.

However, I explained to my gunners the shooting procedure that I wanted to follow, and added that no one was allowed to shoot without my loud and clear approval! In a mine field, mistakes are absolutely and totally unforgiving. We began destroying mines on the north side among the Sampans' debris and the human remains, which were few because most of the body parts had already sunk beneath the waves. We destroyed six mines the first hour, which was not enough if we expected to complete the job before dark. I decided to move to the south side of the mines so that when dusk came they would not drift out of my sight into the darkness before I could devise a plan to keep up with all of them. After that move, we destroyed nine more mines before darkness forced us to cease fire.

"Talker, get your pencil and clipboard, I want to send a message to our ship's captain."

"Aye aye Captain, I'm ready."

"Request permission to remain overnight to monitor the remaining free-floating mines and complete my mission tomorrow. From my last reported position I am moving due south, riding an estimated two-knot current while keeping a close watch on eight unexploded mines. So far on this mission we have already destroyed fourteen Sampans and exploded fifteen free-floating mines. I can only guess how many mines were on board the Sampans that we exploded with our gunfire, but perhaps fifteen to twenty would be a close count. Will complete my assignment tomorrow beginning as soon as we can clearly see those deadly nightmares. Until then I must carefully shepherd the leftovers until daybreak."

Perkins captain answered, "Permission granted Captain, your team has done a remarkable job today, handle the remaining mine situation as you see fit. We'll be standing by until all of you are safely back on board. *Perkins* captain out."

Everyone except my engineer and my talker was given a battle lantern and assigned by me to a mine lookout post.

Chapter 57

Those assignments would be critical for the duration of the night. Since we were in a hot zone, no Russian interference was expected, so we used the lights as we pleased and that became plenty. There were three DC recharge outlets on the engine and they were in constant use. I had brought twelve lanterns, therefore our lighting was sufficient to protect us from colliding with off-course mines.

Every man was told to, without a moment's delay, inform me if there were any sign of a bearing change from the particular mine that he was watching. And then I reminded them all that there is no idle time to pass in a mine field unless you have already reserved a seat in the hereafter. I restricted all talk to an audible whisper except when it became necessary to inform me that we were losing sight of a mine that was not behaving properly (rogue). Then I wanted the informant to HOLLER!

We could not in any way change the course of a mine, all I could do was to move our boat out of its path and we did that continuously throughout the night. However, to perform that getting out of the way maneuver I always backed our boat up-current to maintain maximum control, because going with the flow of a current reduces a captain's control of his vessel by at least fifty percent. If I backed equal to the current's speed plus a little bit, the mine would move safely away from us. Our biggest nightmare would be if two or more mines drifted together and exploded. An event of that kind could possibly set off a chain reaction, depending upon the unseen drift patterns. I thanked God repeatedly for the calm sea we were experiencing. As it was, we danced at death's door from dusk until dawn, about eleven hours.

When sunrise had finally revealed the mines, my entire team looked at me and smiled. No talk, just smiling faces! It was then that I knew they had absolutely accepted me as their new captain. After I had accounted for all eight mines I asked, "Is there any food left?"

"There is but it's cold," I was told.

"Can we heat the coffee on the engine?"

The engineer answered, "Pass it to me I believe that I can."

But as it turned out no one wanted to eat or drink, however they did want to get on with the task of destroying the final eight mines so we could return to our ship, get warm, and eat hot food. How different this team is from my experiences with the Rangers, I thought. How correct I was when I made the duty comparison to the chief corpsman – this is limited duty!

Anyway, I had just begun to consider the best procedure to use while blowing up the final eight when the gunner's mate who had been protecting my back asked, "Captain, can I shoot one of the remaining mines?"

Suddenly, I felt as a father might feel while teaching his son to shoot on their first hunting trip. "You bet you can," I answered. "Are you checked out on the BAR?"

"No, I'm not, but from what I saw yesterday I can hit one of those mines with my M-1 rifle."

I did not tell him that I had used an M-1 rifle many times to blow up mines. "All right sailor, the boat is jumping a bit more than yesterday but I will point you toward my chosen target and stay there until you decide to shoot."

He fired and exploded the selected mine with one shot. Everyone cheered and began calling him Daniel Boone.

"Would you like to try another one?" I asked.

He bobbed his head to indicate that he would. I motored to number two and put him in a very favorable position. That time he took longer to shoot but when he did, one shot exploded another mine. The BAR shooters shouted, "Let him continue Captain!"

I chose another mine and pointed him toward it. That time it took two shots, which was far better than I had ever done. The weather was picking up so I let him continue shooting because turbulent weather is pure hell when working with floating mines, which have only fifty percent bouncy at best.

The remaining mines began to appear and then disappear in the increasingly choppy sea. The final three were very time-consuming to destroy. Therefore, I had ordered all shooters to fire at each appearance of a mine that I had chosen to destroy. After we had exploded number eight, I radioed *Perkins* and requested a fly-over to make sure that I had not lost one. Two hours later I got the answer – "No mines spotted in your area Captain."

"Do I have permission to return to *Perkins*?"

After I had been given permission to do that, a course for that purpose was provided by our navigator. It was late afternoon when I sighted *Perkins* approaching us at flank speed (as fast as she could go). Every off-duty sailor had lined the port side decks watching for our return. We had scored another big combat score for *Perkins*!

When we were about a mile apart, our captain radioed and asked what pick-up speed I wanted.

"Give me ten knots on port side lee, sir."

"Boats your team is not trained for that kind of pick-up!"

"Sir, they will be in about ten minutes."

"All right ten knots it is … good luck."

Both vessels began to line up for the high speed recovery. I did not tell the captain but I had already begun to interest my team in a fantastic recovery during our trip back. "We have had a fantastic mission mates so let's keep it fantastic until we step out of our boat onto *Perkins*' main deck."

Without realizing it I had learned from my Sergeant how to get my men excited about doing whatever I wanted them to do. I had sufficient time to instruct, but the danger zone would be lack of experience. To compensate for that deficiency, I began instructing two of my most experienced sailors on sea painter seamanship. "Get this one right or the others won't matter," I taught. "Talker, get word to the recovery officer that I do not want a painter messenger line used."

"Aye, aye Captain."

I had also instructed the engineer on how to work with me on the approach, and when that time came he did a marvelous job. The green light on the bridge deck came on so I began my close-in approach just forward of the port propeller guard, while focusing on a ten-foot clearance between *Perkins*' steel hull and my wooden hull.

Chapter 58

My engineer matched *Perkins*' hull speed plus a little bit, while I perfectly steered my boat beneath her two large hoisting blocks, so that my sea painter trainee could calmly reach out and be handed, by a member of the deck recovery crew, the eye-spliced painter which he promptly secured to our boat's Samson post. Every witness applauded and cheered loudly for the confident skill which my team had developed. All of the other recovery events were done with equal confidence and style. After we had disembarked on board *Perkins*' main deck and I had surrendered our boat to the recovery officer, the witnesses went wild and welcomed us home. Even though it was a very flattering out-pouring, my time with them was cut short by the chief corpsman (medic) who without delay led me to sickbay for a lengthy physical examination. His medical lecture to me took even longer. But in time he came around to accepting my defense for staying out overnight, then did a flip-flop and began complimenting my team for what we had done for our Navy and our country.

"Well so far so good I guess but I'm still looking forward to your being ordered to San Diego for discharge before you are either injured or killed. Because if either of those misfortunes should happen to you, I will surely be sent to San Diego and discharged, but as a seaman apprentice who had ignored orders from a superior Navy medical officer. Also my retirement pension would be completely in the toilet."

"Now Chief, don't fret because when that last mission's battle score has been reported to the admiral, you might even be decorated for allowing me to go."

"Was it that big?"

"Yes, it was Chief!"

"Thank God, maybe I can sleep tonight."

The chief and I left sickbay and proceeded to the mess deck where a big feed had been prepared for my team. After I had rejoined them, accolades continued uninterrupted for five minutes or more, but finally faded away so that a very hungry group of young sailors could eat and drink their fill. Present team members had been allowed to begin eating while I was having a medical check. However, it didn't take long for me to catch up.

During my absence, our captain told *Perkins'* crew as much as he was allowed about my nine month's assignment with the U.S. Army Rangers. He followed that with, "Now he is back home onboard *Perkins*. It is my intention to have him safely (free from KGB) delivered to the separation center in San Diego on June first or as close to that date as possible. In the meantime he will serve as my assault coxswain and here is why! On the mission that his team has just completed they ...." He read to them our battle record from our most recent mission.

"If my information is correct, Boats will be with us for six more weeks."

The chief medic who was sitting next to me at a dinning table whispered, "Oh Lord, please help me!" We looked at each other and smiled.

During my remaining time on board *Perkins*, even though I would complete twenty-eight more missions, the chief and I became very close friends. The floating mine issue had become a double-edged sword and there seemed to be an ever-increasing number of them. The enemy's new tactic was to deploy larger Sampans which were armed with better weapons and fully rigged for offshore (blue water) fishing as a disguise.

But the truth was they carried both mines and troops, troops which would infiltrate Korea south of the thirty-eighth parallel – behind UN lines. Their favorite ploy, after being confronted, was to hold up a very young child, to send a message to the attacker that there were children on board their imperiled vessel. That trap had taken the lives of many assault boat sailors who believed the Sampan they were about to attack was no more than an innocent family, fishing to live. What a deadly mistake that turned out to be!

Therefore, we also changed our methods of attack. For example, if there were any doubt, we never approached closer than one thousand yards (one half mile), with all BARs at the ready, and then radioed for air cover support. A British aircraft carrier (HMS *Ark Royal*) had been assigned as the aircraft support vessel for my patrol area.

I will write about only one of those final twenty-eight missions because it's the one, even to this day, that remains in my troubled dreams. While approaching a large Sampan which had strayed into my patrol area, I ordered my engineer to heave to (stop) because I had become very uncertain about my potential target's threat status. Therefore, I positioned my boat out of range of any weapon it might have had on board and called for air cover support. For reasons unknown to me I had not been able to make radio contact with any airplane. After waiting about thirty minutes without any contact with an aircraft, I ordered a red signal flare to be fired straight up to make it easier for the Ark Royal's pilot to locate us.

God only knows why, but here I've written the sad consequences of that tragic failure to communicate. He (pilot) had apparently seen my flare because very soon after it had been fired, his combustion engine Hurricane fighter plane zoomed by us. We all waved to the pilot. He rocked his wings, rolled over into a dive, and leveled off about fifty feet above the water. After making sure that we were clear, the pilot began firing fifty-caliber bullets at the defenseless Sampan and in seconds it was totally destroyed with all on board. There was nothing left of it but splinters and other flotsam.

I had called for cover protection, NOT for destroy; but what happened was absolutely caused by my not being able to communicate with the British pilot. There had been no mine explosions and no bullets fired at us from the targeted Sampan so those people truly were innocent fishermen. For whatever reason, I then made a big mistake – I maneuvered my boat through the Sampan's debris from the mistaken air attack.

Chapter 59

There are things I wish I had not seen that day because what happened still haunts me in my dreams – the dead bodies of three Korean adults and three children, but no enemy troops, are indelibly imprinted on my mind and will be for as long as I live. This time it was not a ploy, I grieved.

My captain, who had highly praised my performance while serving my country under his command, released me from any further military duty on board *Perkins*. I was back in the custody of Secret Service people. With two of them (SS) who had been brought aboard *Perkins* by a U.S. Navy helicopter, I waited several hours for a U.S. Navy Seaplane (PBY) which had been commandeered by the Secret Service to land on the Sea of Japan, weather permitting, and ferry us to Tokyo Bay.

Once again I had absolutely no privacy but that really didn't matter much anymore, even though the Russian KGB, as I was told, had stepped up its efforts to either kill or capture me. During that long wait I had plenty of time to express farewells to many shipmates whom I would sadly never see again. With that knowledge it was a much more emotional thing to do, but sooner or later all things must come to an end, I thought.

The whaleboat had been decorated with many colorful pennants to honor what I had done while serving as *Perkins'* assault coxswain. Also, I was asked to coxswain her to the waiting PBY, which would prove to be my last time at the tiller. That request got to me and I could not hold back the tears. I just stopped trying and let them flow. It was especially emotional when my boat was, for the last time, with me as her captain, slowly lowered onto the sea.

My ex-captain saluted and waved goodbye to me from the bridge deck. I returned his salutation and mumbled to myself, "Goodbye sir." The already-landed PBY was experiencing considerable difficulty holding position on the windy Sea of Japan. I told my talker to ask the pilot to maintain a course and speed that would hold his airplane into the wind, and then pass to us a thirty-foot tow line (painter). The propeller wind blast made that very hard for us to do, even with considerable help from our boat's engine, but my idea kept us clear of the deadly propellers and made it much easier when the time came for the pilot's crew to help pull us aboard his aircraft.

After the two Secret Service people and I were safely on board the seaplane, the bow man released the painter from the whaleboat's Samson post and that would turn out to be the last time in my life that I would experience that very familiar exercise. From Tokyo we continued to Midway Island on a large military transport plane, then on the same aircraft to Pearl Harbor, and finally to San Diego, California, arriving on May 31, 1952, at three a.m.

At last, by the miraculous grace of God, I was safely back in my own country among my own people. When our plane landed at San Diego Municipal Airport, there was a U.S. Navy automobile waiting for it. From there I was taken to the U.S. Navy's west coast separation center at their large destroyer base, where we were housed in guarded quarters.

Over the coming days, I was reminded over and over how important it was for me to be protected from captivity or death. "Because if you are captured by KGB," I was told, "our debating strategy at the UN, which is to continue accusing Russia and China of being very much involved in the invasion of South Korea, will be completely futile. On the other hand if you are killed, our debating position at the UN will also be adversely affected because after that tragedy, we could not produce the witness we have claimed to have. KGB well knows from evidence found on board your assault boat which was left at the fish factory that you were its captain. Therefore, you are their primary target. However, that does not mean they are not looking for the other members of your assault team, because believe me they are!

"Since Russia's international reputation has been compromised, they will not give up the hunt easily. Consequently, neither will we give up protecting you. We have purged all of your military and civilian records so they (KGB) will not be led to your doorstep. Now all of this may sound very difficult but actually it is not because all you have to do is KEEP YOUR MOUTH SHUT! You should not talk with anyone about your service with the Army Rangers and that is final."

"For how long?" I asked.

"I don't know the answer to that, Boats."

They spent days with me in and around San Diego, teaching me how to be a civilian again and how to recognize conversation that could lead to trouble. Also I was taught how to courteously remove myself from any conversation that might lead to talking about my military experience. They bought second-hand civilian clothes and shoes for me to wear on the train ride back to my home. Those clothes were anything but attractive; however, that was exactly what they wanted. I was warned not to become friendly with anyone during my trip home.

After three days of required medical necessities and paperwork, and having been honorably discharged from the United States Navy on June sixteenth, 1952, I walked back to our quarters with the two SS men and was without delay driven to Los Angeles Union Station (railway).

During that drive they told me something that really got my attention. "If you cannot protect yourself we will do it for you!"

"Such as?" I inquired.

"We will put you someplace where we can protect you."

"For example?" I asked.

"We have a witness protection program. But as I have already told you, just keep your mouth shut about your military record until the Korean conflict is over and then, there will more than likely be no more problems with that country. Therefore, you will not be needed."

Chapter 60

However, no one back then could have predicted that Korea would be the longest war in our country's history! It's been going on fifty-five years or more and it ain't over yet! The high energy debate at the UN between Russia and the USA concerning who was responsible for starting the Korean War was instantly upstaged by an object of worldwide interest called "Sputnik."

That completely unprecedented technical step into space forced both of the debater's to refocus their attention on a small satellite, about the size of a basketball, in outer space. That space leap forward by Russia became a serious new challenge for USA because it quickly captured the eyes of the whole world, and to some extent created fear. And Mother Russia, with its frightening lead in space technology, along with the death of Joseph Stalin, quickly regained her prestigious reputation as a world superpower. Anyway, after Sputnik, I became very low profile and seldom heard from the Secret Service. When I came back on their radar, I requested permission to write this book, and all I could get from them (SS) was a big NO. But I persevered and finally, after years of trying, got everything I wanted except one important thing. I was told that I could not use anyone's name in my book, not even my own. That restriction caused continued writing to be very difficult, but once again I persevered and was given permission to use my name only, as author on the book's cover.

In this book, I have been given permission to write about an undisclosed history of the Korean War in which I participated, and have now made it available to anyone who would like to learn about it. Also from that war, I have in this book shared with the reader frequent nightmares that have haunted my dreams for more than half a century.

THE END.

Epilogue

Throughout the years, I have often thought of my wartime brothers (assault-boat team mates) who, with me, had willingly risked their lives a dozen times over while proudly serving our beloved country. I have wondered many times if they had survived their youthful military experience, if they had married and if they had fathered children or perhaps grown old - like me. A few years ago I made a special trip to Ashville, North Carolina in search of Sam, even though I was clearly aware that I had been sworn not to attempt in any way to contact my disbanded team mates. But after thirty years, the urge to locate my best friend and brother overwhelmed me. After several days of searching and calling everyone in Buncombe county who shared his last name, I struck out. Absolutely no one with his family name, even though in the county where he grew up, would admit to having ever known him. With all of the others almost the same thing had happened. It was as if they had never existed or worse yet, as if they were figments of my imagination. However, each of my brothers still lives in my memory of what we had done together in North Korea and Russia and those treasured memories will remain with me until I cross the bar and join them in blissful eternity.

William C. Warwick, Jr. received this decoration from the government of South Korea fifty years after his service to that country.

Image: Elizabeth Warwick